Applications of NMR Spectroscopy
(Volume 8)

Edited by
Atta-ur-Rahman, *FRS*
Kings College
University of Cambridge
Cambridge
UK

&

M. Iqbal Choudhary
H.E.J. Research Institute of Chemistry,
International Center for Chemical and Biological Sciences,
University of Karachi, Karachi,
Pakistan

Applications of NMR Spectroscopy

Volume # 8.

Editors: Atta-ur-Rahman, *FRS* and M. Iqbal Choudhary

ISSN (Online): 2405-4682

ISSN (Print): 2405-4674

ISBN (Online): 978-981-14-3997-1

ISBN (Print): 978-981-14-3383-2

ISBN (Paperback): 978-981-14-3384-9

©2020, Bentham Books imprint.

Published by Bentham Science Publishers Pte. Ltd. Singapore. All Rights Reserved.

BENTHAM SCIENCE PUBLISHERS LTD.
End User License Agreement (for non-institutional, personal use)

This is an agreement between you and Bentham Science Publishers Ltd. Please read this License Agreement carefully before using the book/echapter/ejournal (**"Work"**). Your use of the Work constitutes your agreement to the terms and conditions set forth in this License Agreement. If you do not agree to these terms and conditions then you should not use the Work.

Bentham Science Publishers agrees to grant you a non-exclusive, non-transferable limited license to use the Work subject to and in accordance with the following terms and conditions. This License Agreement is for non-library, personal use only. For a library / institutional / multi user license in respect of the Work, please contact: permission@benthamscience.net.

Usage Rules:

1. All rights reserved: The Work is the subject of copyright and Bentham Science Publishers either owns the Work (and the copyright in it) or is licensed to distribute the Work. You shall not copy, reproduce, modify, remove, delete, augment, add to, publish, transmit, sell, resell, create derivative works from, or in any way exploit the Work or make the Work available for others to do any of the same, in any form or by any means, in whole or in part, in each case without the prior written permission of Bentham Science Publishers, unless stated otherwise in this License Agreement.
2. You may download a copy of the Work on one occasion to one personal computer (including tablet, laptop, desktop, or other such devices). You may make one back-up copy of the Work to avoid losing it.
3. The unauthorised use or distribution of copyrighted or other proprietary content is illegal and could subject you to liability for substantial money damages. You will be liable for any damage resulting from your misuse of the Work or any violation of this License Agreement, including any infringement by you of copyrights or proprietary rights.

Disclaimer:

Bentham Science Publishers does not guarantee that the information in the Work is error-free, or warrant that it will meet your requirements or that access to the Work will be uninterrupted or error-free. The Work is provided "as is" without warranty of any kind, either express or implied or statutory, including, without limitation, implied warranties of merchantability and fitness for a particular purpose. The entire risk as to the results and performance of the Work is assumed by you. No responsibility is assumed by Bentham Science Publishers, its staff, editors and/or authors for any injury and/or damage to persons or property as a matter of products liability, negligence or otherwise, or from any use or operation of any methods, products instruction, advertisements or ideas contained in the Work.

Limitation of Liability:

In no event will Bentham Science Publishers, its staff, editors and/or authors, be liable for any damages, including, without limitation, special, incidental and/or consequential damages and/or damages for lost data and/or profits arising out of (whether directly or indirectly) the use or inability to use the Work. The entire liability of Bentham Science Publishers shall be limited to the amount actually paid by you for the Work.

General:

1. Any dispute or claim arising out of or in connection with this License Agreement or the Work (including non-contractual disputes or claims) will be governed by and construed in accordance with the laws of Singapore. Each party agrees that the courts of the state of Singapore shall have exclusive jurisdiction to settle any dispute or claim arising out of or in connection with this License Agreement or the Work (including non-contractual disputes or claims).
2. Your rights under this License Agreement will automatically terminate without notice and without the

need for a court order if at any point you breach any terms of this License Agreement. In no event will any delay or failure by Bentham Science Publishers in enforcing your compliance with this License Agreement constitute a waiver of any of its rights.
3. You acknowledge that you have read this License Agreement, and agree to be bound by its terms and conditions. To the extent that any other terms and conditions presented on any website of Bentham Science Publishers conflict with, or are inconsistent with, the terms and conditions set out in this License Agreement, you acknowledge that the terms and conditions set out in this License Agreement shall prevail.

Bentham Science Publishers Pte. Ltd.
80 Robinson Road #02-00
Singapore 068898
Singapore
Email: subscriptions@benthamscience.net

CONTENTS

PREFACE	i
LIST OF CONTRIBUTORS	ii

CHAPTER 1 qNMR AS A TOOL FOR DETERMINATION OF SIX COMMON SUGARS IN FOODS 1
Wen-Bin Yang, Shu-Huey Wang and *Yi-Ting Chen*

INTRODUCTION	1
RESULTS	3
Workflow 1: Measurement of 6 Common Sugars in Foods	3
Sample Preparation	3
NMR Experimental Process	3
Statistical Analysis	4
NMR Spectral Analysis of Six Common Sugars in Beverages and Foods	5
NMR Spectral Analysis of Mixed Sugars in Foods	8
Introduction of NAIM Labeled Sugar	11
General Procedure for Preparation of Sugar-NAIM Derivatives	12
NAIM Derivatization and NMR Spectrometric Data of Aldo-Sugars	12
NMR Spectrometric Analysis of Mixed Sugars via NAIM Derivatization	14
Workflow 2: Measurement NAIM Labeled Sugars, Fru and Suc in Foods	15
Sample Preparation	16
NMR Experimental Process	17
Statistical Analysis	17
DISCUSSION	20
CONCLUSIONS	20
ABBREVIATIONS	21
CONSENT FOR PUBLICATION	22
CONFLICT OF INTEREST	22
ACKNOWLEDGEMENTS	22
REFERENCES	22

CHAPTER 2 CORRELATION BETWEEN VIP SCORES AND 1H NMR TO EXTRACT INFORMATION OF PSYCHOLOGICAL ATTENTION TESTS APPLIED BEFORE AND AFTER COFFEE INTAKE 25
Michel Rocha Baqueta, Aline Coqueiro, Letícia de Sousa Frutuozo, Paulo Henrique Março, Frank Duarte, Manuela Mandrone, Ferruccio Poli and *Patrícia Valderrama*

INTRODUCTION	26
EXPERIMENTAL	28
Psychological Tests and Coffee Intake	28
Sample Extraction for ^{1}H NMR Analysis	30
^{1}H NMR Analysis	30
Chemometrics	31
RESULTS AND DISCUSSIONS	32
Human Attention and Coffee Preference	32
Correlation between the Coffee Chemical Profile and Psychological Tests	35
CONCLUDING REMARKS	38
CONSENT FOR PUBLICATION	38
CONFLICT OF INTEREST	38
ACKNOWLEDGEMENTS	38
REFERENCES	39

CHAPTER 3 NMR SPECTROSCOPY FOR PROBING THE STRUCTURAL DETERMINANTS OF APTAMER OPTIMIZATION AND RIBOSWITCH ENGINEERING 42
B. Bora1, Ö. Uğurlu, E. Man1, M. Gültan, C. Özyurt and S. Evran
   INTRODUCTION 43
      Selection of Aptamers 43
      Post-SELEX Modifications of Aptamers 44
      Riboswitches 46
   NMR AND COMPLEMENTARY TECHNIQUES FOR APTAMERS AND RIBOSWITCHES 48
   CONCLUSION 53
   CONSENT FOR PUBLICATION 53
   CONFLICT OF INTEREST 53
   ACKNOWLEDGEMENTS 53
   REFERENCES 54

CHAPTER 4 APPLICATIONS OF NMR SPECTROSCOPY IN MEDICAL DIAGNOSIS 61
Baharudin Ibrahim and Keshamalini Gopalsamy
   INTRODUCTION 62
   WORKING PRINCIPLE OF NMR IN MEDICAL DIAGNOSIS 62
      NMR in the Diagnosis of Lung Cancer 63
         Background of Lung Cancer 63
         Study by Carrola et al. (2011) 64
      NMR in the Diagnosis of Alcohol Use Disorder (AUD) 69
         Background of Alcohol Use Disorder (AUD) 69
         Study by Mostafa et al. (2016, 2017) 70
      NMR in the Diagnosis of Parkinson's Disease 73
         Background of Parkinson's Disease 73
         Study by Ahmed et al. (2009) 73
      NMR in the Diagnosis of Other Diseases 76
   CONCLUSION 76
   CONSENT FOR PUBLICATION 76
   CONFLICT OF INTEREST 77
   ACKNOWLEDGEMENTS 77
   REFERENCES 77

CHAPTER 5 APPLICATIONS OF NMR SPECTROSCOPY IN CANCER DIAGNOSIS 81
Asmaa A. Kamel and Fotouh R. Mansour
   INTRODUCTION 82
   OVERVIEW OF NMR SPECTROSCOPY 83
      Types of NMR Spectroscopy Used in Cancer Diagnosis 83
      Advantages and Disadvantages of NMR Spectroscopy 84
   TECHNICAL ASPECTS OF IN VITRO NMR APPLICATIONS 86
   APPLICATIONS 87
      Brain Tumor 87
      Breast Cancer 88
      Ovarian and Endometrial Cancer 91
      Prostate Cancer 93
      Lung Cancer 96
      Colorectal Cancer 101
      Urinary Bladder Cancer 106
      Oral Cancer 107

PERSPECTIVE AND CONCLUDING REMARKS	109
CONSENT FOR PUBLICATION	110
CONFLICT OF INTEREST	110
ACKNOWLEDGEMENTS	110
REFERENCES	110

CHAPTER 6 NMR AS A TOOL FOR EXPLORING PROTEIN INTERACTIONS AND DYNAMICS ... 121

Qamar Bashir and *Naeem Rashid*

INTRODUCTION	122
Protein Dynamics and the Encounter Complex	123
Chemical Shift Perturbation Analysis	124
Target Immobilized NMR Screening	127
Paramagnetic Relaxation Enhancement	129
CONCLUDING REMARKS	132
CONSENT FOR PUBLICATION	133
CONFLICT OF INTEREST	133
ACKNOWLEDGEMENTS	133
REFERENCES	133
SUBJECT INDEX	141

PREFACE

Nuclear Magnetic Resonance (NMR) spectroscopy has emerged as one of the most powerful techniques for the identification of materials, and for the study of their dynamic properties. As a result, the technique has found tremendous uses in almost all fields of physical, natural, and health sciences.

Volume 8 of the book series entitled *Applications of NMR Spectroscopy* is mainly focussed on the practical uses of NMR spectroscopy in solving various key problems in biomedical, health, and food sciences. The contents include NMR based analysis of common sugars, plant based constituents, nucleic acids and proteins, as well as NMR- based metabolomics and MRI for the diagnosis of chronic and acute health disorders.

The review contributed by Yang *et al.* provides an overview of the use of quantitative NMR (qNMR) techniques for the analysis of six common sugars in complex food matrices, after derivatization with naphthimidazole (NAIM). Coffee plants contain many constituents which effect cognitive functions. Valderrama *et al.* have analysed the constituents of various coffee types, and correlated them through the use of the psychological attention test. Evran *et al.* have reviewed the recent literature on the NMR-based structures of aptamers (single stranded DNA and RNA molecules) selected via an iterative process called Systematic Evolution of Ligands by Exponential Enrichment (SELEX). The applications of NMR for the identification of the ligand binding mechanisms are discussed. The review contributed by Ibrahim and Gopalsamy describes various NMR techniques used in metabolomics-based diagnosis of common diseases. The next chapter by Asmaa and Mansour provides a critical analysis of various MRI-based diagnostic approaches to study diverse human cancers. Proteins are fascinating molecules, both because of their complex structures and their interactions with other biomolecules. NMR techniques have evolved over the years to determine the structures and functions of protein molecules. This is the key theme of chapter 6 by Bashir and Rashid.

We wish to thank all the eminent scientists for their scholarly contributions. The editorial team of Bentham Science Publishers, particularly Ms. Fariya Zulfiqar (Manager Publications) and team leader Mr. Mahmood Alam (Director Publications), deserves our deepest appreciation for compiling an excellent volume in a time efficient manner. We are confident that like the previous volumes of this book series, the current treatise will also receive wide appreciation for both the readers and practitioners of NMR spectroscopy.

Prof. Dr. Atta-ur-Rahman, *FRS*
Honorary Life Fellow
Kings College
University of Cambridge
Cambridge
UK

&

Prof. Dr. M. Iqbal Choudhary
H.E.J. Research Institute of Chemistry
International Center for Chemical and Biological Sciences
University of Karachi
Karachi
Pakistan

List of Contributors

Aline Coqueiro	Universidade Tecnológica Federal do Paraná, Ponta Grossa, Paraná, Brazil
Asmaa A. Kamel	Biochemistry Department, Faculty of Pharmacy, Tanta University, Tanta 31111, Egypt
B. Bora	Department of Biochemistry, Faculty of Science, Ege University, 35100 Bornova-İzmir, Turkey
Baharudin Ibrahim	School of Pharmaceutical Sciences, Universiti Sains Malaysia, Penang, Malaysia
C. Özyurt	Department of Chemistry and Chemical Processing Technologies, Lapseki Vocational School, Canakkale Onsekiz Mart University, Canakkale, Lapseki, Turkey
E. Man	Department of Biochemistry, Faculty of Science, Ege University, 35100 Bornova-İzmir, Turkey
Fotouh R. Mansour	Department of Pharmaceutical Analytical Chemistry, Faculty of Pharmacy, Tanta University, Tanta 31111, Egypt Pharmaceutical Services Center, Faculty of Pharmacy, Tanta University, Tanta 31111, Egypt
Ferruccio Poli	Università di Bologna, Bologna, Italy
Frank Duarte	Faculdade União de Campo Mourão, Campo Mourão, Paraná, Brazil
Keshamalini Gopalsamy	School of Pharmaceutical Sciences, Universiti Sains Malaysia, Penang, Malaysia
Letícia de Sousa Frutuozo	Universidade Tecnológica Federal do Paraná, Campo Mourão, Paraná, Brazil
Michel Rocha Baqueta	Universidade Tecnológica Federal do Paraná, Campo Mourão, Paraná, Brazil
M. Gültan	Department of Biochemistry, Faculty of Science, Ege University, 35100 Bornova-İzmir, Turkey
Manuela Mandrone	Università di Bologna, Bologna, Italy
Naeem Rashid	School of Biological Sciences, University of the Punjab, Quaid-e-Azam Campus, Lahore 54590, Pakistan
Ö. Uğurlu	Department of Biochemistry, Faculty of Science, Ege University, 35100 Bornova-İzmir, Turkey
Patrícia Valderrama	Universidade Tecnológica Federal do Paraná, Campo Mourão, Paraná, Brazil
Paulo Henrique Março	Universidade Tecnológica Federal do Paraná, Campo Mourão, Paraná, Brazil
Qamar Bashir	School of Biological Sciences, University of the Punjab, Quaid-e-Azam Campus, Lahore 54590, Pakistan
Shu-Huey Wang	Core Facility Center, Department of Biochemistry, Taipei Medical University, Taipei 110, Taiwan R.O.C
S. Evran	Department of Biochemistry, Faculty of Science, Ege University, 35100 Bornova-İzmir, Turkey

Wen-Bin Yang	The Glycan Sequencing Core Facility, Genomics Research Center, Academia Sinica, Taipei 115, Taiwan R.O.C
Yi-Ting Chen	The Glycan Sequencing Core Facility, Genomics Research Center, Academia Sinica, Taipei 115, Taiwan R.O.C
Ö. Uğurlu	Department of Biochemistry, Faculty of Science, Ege University, 35100 Bornova-İzmir, Turkey

CHAPTER 1

qNMR as a Tool for Determination of Six Common Sugars in Foods

Wen-Bin Yang[1,*], Shu-Huey Wang[2] and Yi-Ting Chen[1]

[1] *The Glycan Sequencing Core Facility, Genomics Research Center, Academia Sinica, Taipei 115, Taiwan R.O.C.*

[2] *Core Facility Center, Department of Biochemistry, Taipei Medical University, Taipei 110, Taiwan R.O.C.*

Abstract: Nuclear magnetic resonance (NMR) spectroscopy is capable of quantifying molecules. The term so called quantitative NMR (qNMR), has been used for determination of the concentration and purity of small molecules. Carbohydrates are found in various beverages and dietary foods, including crops, milk, fruits, and vegetables. Commercial products frequently use "added sugar" in soft drinks, cookies, candies, and foods. The added sugar in beverages can be sucrose, high-fructose corn syrup (HFCS) and glucose. Here, we report a quantitative method to measure 6 common sugar ingredients in foods from a single one-dimensional ^{1}H-NMR and by using naphthimidazole (NAIM) derived sugars, which are chemically tagging aldoses with 2,3-naphthalenediamine (NADA) at the reducing ends to assist assignment of sugars. The aldoses in native sugars contain α and β anomeric isomers, and may have overlapping signals in ^{1}H-NMR spectra. In contrast, both the anomeric isomers can be converted into a single sugar-NAIM derivative, which resolves the problem of overlapping signals to simplify the NMR quantitative analysis. This NAIM method is especially useful for identification and quantification of multiple kinds of sugars in beverages and foods. This study is to facilitate the quantification of six common sugars in beverages and foods. Our results suggest that a simple treatment of beverage and food with the NAIM labeling method provides a more extensive success rate for the quantification of sugar ingredients.

Keywords: Beverage, Food, Fructose, Galactose, Glucose, Lactose, Maltose, Naphthimidazole (NAIM), q-NMR, Quantitative analysis, Sugar, Sucrose, ^{1}H-NMR spectrometry.

INTRODUCTION

Carbohydrates are found in various beverages and dietary foods, including rice, noodles, bread, meat, milk, fruit, vegetables, and drink [1, 2]. Carbohydrates are

---

* **Corresponding author Wen-Bin Yang:** The Glycan Sequencing Core Facility, Genomics Research Center, Academia Sinica, Taipei 115, Taiwan R.O.C.; Tel: +886-2-27871264; E-mail: wbyang@gate.sinica.edu.tw

Atta-ur-Rahman and M. Iqbal Choudhary (Eds.)
All rights reserved-© 2020 Bentham Science Publishers

also used as "added sugar" in soft-drinks, cookies, candies, and many kinds of foods. For example, the added sugar in beverages can be sucrose, fructose, glucose, maltose and other sweeteners. Though carbohydrates are needed for living, an excessive uptake of sugar may induce health problems such as decayed teeth and chronic diseases [3 - 5]. In addition, foods of low glycemic index (GI) are suggested for diabetic patients. It is important to know the content and quantity of sugar in foods. Thus, developing a rapid and convenient qualitative/quantitative method for sugar measurement in foods is needed. Furthermore, many countries have introduced the sugar tax and soft-drink tax in order to reduce sugar consumption [6]. Therefore, a suitable method to verify the sugar content in foods can be provided to the government for policy implementation. The appropriate "fine sugar" or "added sugar" intake is 25 grams per day according to the scientific recommendation by the World Health Organization (WHO) [7]. Since August 2015, Taiwan Food & Drug Administration (TFDA) has proposed to regulate common sugars in foods, including glucose (Glc), galactose (Gal), fructose (Fru), lactose (Lac), maltose (Mal), and sucrose (Suc). The amounts of sugars must be labeled in the "Nutrition Facts Panel" for the products of beverages and foods. Even though the information of sugar content surely benefits consumers, this regulation will impose challenges to the food industry concerning the identification and quantification of the six common sugars in beverages and foods.

At present, high performance liquid chromatography (HPLC) and high-performance anion-exchange chromatography with pulsed amperometric detection (HPAEC-PAD) are more common instrumental methods for sugar determination in foods. NMR spectroscopy is also a powerful method for identification and quantification of low molecular weight compounds. Though ^{1}H-NMR spectra are commonly used in the routine quantitative analysis of individual sugars [8, 9], using NMR to identify each sugar in a mixture and simultaneously quantify its content is still challenging because the spectrum is usually complicated by the existence of anomeric isomers and by the similar structures of sugar components. The quantitative NMR (qNMR) technique is designed for determination of the concentration and purity of small molecules [10]. qNMR can be applied for direct quantification of multiple components in a mixture without pretreatment of sample. However, recording a qNMR spectrum would take a much longer acquisition time than a routine ^{1}H-NMR spectrum. In another approach, we performed a simple treatment on beverages and foods with a naphthimidazole (NAIM) labeling kit to provide the sugar-NAIM derivatives for quantification by ^{1}H-NMR spectral analysis. This method combining NAIM derivatization and NMR analysis is successfully applied to the measurement of six common sugars in foods. Our objective is to establish a convenient method for profiling and quantifying sugar ingredients in beverages and foods by using one-dimensional

^{1}H-NMR spectroscopy *via* a simple treatment with NAIM labeling kit.

RESULTS

Workflow 1: Measurement of 6 Common Sugars in Foods

Using ^{1}H-NMR for six common sugars (Glc, Gal, Fru, Suc, Mal, and Lac), the identification process was followed stepwise by sample preparation, NMR processing and statistical analysis. Fig. (**1**) shows the flowchart.

Workflow

```
Sample Preparation  →  Solution-state sample          Solid-state/Pasty sample
                       take 50 µL, directly           (1.0 g dissolved in 10 mL H₂O)
                                                      take 50 µL
        ↓
NMR Experimental Process → ¹H NMR/D₂O (1.0 mL)
                            with 0.03% DMSO
        ↓
Statisrical Analysis   →   Calculating 6 sugar content in foods
                           (g/100 mL; g/100 g sample)
```

Fig. (1). Workflow of using ^{1}H-NMR for determination of six common sugars.

Sample Preparation

Six standard sugar solutions (Glc, Gal, Fru, Suc, Mal, and Lac) were prepared in varied concentrations using 5.0, 2.5, 1.25 and 0.25 mg, respectively. The samples of beverage and food in solution-state were ready for determination without pretreatment or separation. A less than 50 µL of sample solution was directly taken to reduce the absorption of H$_2$O (at 4.8 ppm) in ^{1}H-NMR spectra. For solid-state or paste samples, 1.0 gram was dissolved in 10.0 mL of H$_2$O, and 50 µL of the solution was taken for measurement. Sample solution was concentrated in vacuum (3 min), and then deuterium solution was added for NMR experiment.

NMR Experimental Process

The deuteriated water (D$_2$O, 99.9%, Sigma Aldrich, USA) 1.0 mL with 0.03 mol% of dimethylsulfoxide (DMSO, 99.9%, extra dry, H$_2$O < 50 ppm, Acros, New Jersey, USA) as internal standard was added to a dried sample in 5 mm NMR tube for recording the ^{1}H-NMR spectrum. The ^{1}H-NMR spectra were recorded on a Bruker AV600 MHz NMR spectrometer (Bruker BioSpin GmbH,

Rheinstetten, Germany) with a 5 mm dual cryoprobe DCI ^{1}H/^{13}C. Quantification of sugars was based on the integration areas of the characteristic proton signals (e.g. C_1-H in Glc, the α-anomeric H showing a doublet with $J = 6.4$ Hz at 5.20 ppm, and the β-anomeric H showing doublet with $J = 9.8$ Hz at 4.61 ppm) by comparison with that of DMSO (integration region from δ 2.79 to 2.73 ppm for six protons of the two methyl groups). ^{1}H-NMR acquisition parameters: 90° pulse, P1 = 9.95 μs, PL1 = −0.8 dB; relaxation delay D1 = 2 sec; number of acquisition aq = 1.9530824 (s); type of baseline correction: quad; window function: EM; LB = 0.5 Hz; software for spectral processing and regression analysis: TopSpin 3.0.

Statistical Analysis

Calculation of 6 common sugars in foods is based on the integration areas of ^{1}H-NMR spectra. In general, HDO signal is fixed at 4.80 ppm and the integration area of DMSO (δ 2.79 to 2.73 ppm for six protons of the two methyl groups) was set at 6.00. The corresponding peaks of 6 common sugars are shown in Table **1**. In addition, we also use a coefficient K to facilitate the calculation of 6 common sugars in foods. In brief, using qNMR concept the molar proportion of integration areas between DMSO (0.03% in 1.0 mL of D_2O) and sugars (Glc, Fru and Gal have molecular weight 180 g/mol; Suc, Mal and Lac have molecular weight 342 g/mol) were calculated to converse to a unit of g sugar/100 mL or g sugar/100 g in food. To simplify the calculation, we provide a formula as follows for quantification of sugar in foods.

The content of six common sugars in foods (in gram/100 mL or gram/100 g) = K (sugar coefficient) * IA (integration area of sugar in ^{1}H-NMR) * D (times of dilution from original sample/food, i.e., solution-state sample = 1.000; solid-state or paste samples = 10.000).

Table 1. The ^{1}H-NMR integration regions and coefficient K of 6 common sugars.

Sugar Type	Integral Regions (δ, ppm)	Coefficient (K)
Glc	5.26 + 4.47	0.8
Gal	5.29 + 4.60	0.8
Fru	4.13	1.5
Suc	4.23	1.5
Mal	5.44	1.5
Lac	4.48	1.5

The empirical equation (coefficient K) is to facilitate quantification of each common sugar by the integration area of the selected proton signals.

Fig. (2) shows the ^{1}H-NMR spectra of six common sugars in D$_2$O solution (1.0 mL) containing 0.03% DMSO as internal standard. At anomeric center, the C$_1$-H (α- and β-isomers) of Glc (at 5.26 and 4.47 ppm) and Gal (at 5.29 and 4.60 ppm) was calculated as one proton (1 H) in the integration region (Table **1**). In comparison, Fru at 4.13 ppm has 0.5 H in the integration region, whereas disaccharides Mal (at 5.44 ppm), Lac (at 4.48 ppm) and Suc (at 4.23 ppm) respectively have 1 H in the integration region. The signal of HDO was set at 4.80 ppm. The signal of (CH$_3$)$_2$SO was set 6 H in the integration region of 2.79−2.73 ppm.

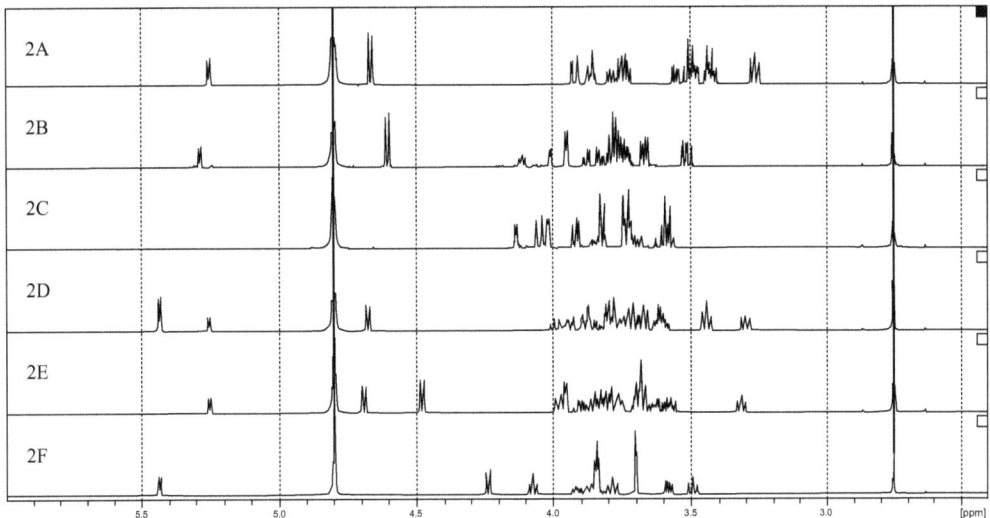

Fig. (2). ^{1}H NMR spectra of six common sugars in D$_2$O solution (1.0 mL) containing 0.03% (CH$_3$)$_2$SO as internal standard: (A) Glc, (B) Gal, (C) Fru, (D) Mal, (E) Lac, and (F) Suc. The signal of HDO was set at δ 4.80 ppm, and the signal of (CH$_3$)$_2$SO occurred at δ 2.73 ppm.

NMR Spectral Analysis of Six Common Sugars in Beverages and Foods

We first examined the ^{1}H-NMR spectrum of six common sugars (Fig. **2**). Taking the integration areas of the characteristic proton signals, one can calculate the amount of each sugar. We used qNMR method and statistical analysis to calculate 6 common sugars in foods. The results are shown in Table **2**. For examples, in the solution-state samples, orange juice has 10.2 g sugar/100 mL comprising Glc (4.1 g), Fru (5.7 g) and Suc (0.4 g); Yogurt milk 2 has 9.2 g sugar/100 mL comprising Glc (2.0 g), Gal (1.7 g), Suc (5.3 g) and Lac (0.2 g); Red wine 2 (wine beverages in Taiwan is no longer to label sugar in "Nutrition Facts Panel", but uses other sweeten labels) has 1.2 g sugar/100 mL comprising Glc (0.6 g) and Fru (0.6 g). In solid-state/paste food samples, potato chip has 1.2 g sugar/100 g comprising Glc (0.5 g) and Suc (0.7 g), chocolate candy has 65.8 g sugar/100 g comprising Suc (56.9 g) and Lac (8.9 g), and honey has 74.1 g sugar/100 g comprising Glc (39.1 g) and Fru (35.0 g), respectively.

Table 2. Quantitative analysis of six individual sugars in commercial beverages and foods by qNMR.

Beverage/Food	Sugar Content in Foods (%, g/100 mL; g/100 g Sample)					
	Glc	Gal	Fru	Suc	Mal	Lac
Winter melon drink[a]	0.1	–	–	7.4	–	–
Lemon beverage	0.3	–	0.3	5.3	–	–
Lemon black tea	3.3	–	4.0	1.2	–	–
Honey-plum vinegar	3.0	–	2.9	7.7	–	–
Refreshment beverage 1	2.8	–	3.1	10.9	–	–
Refreshment beverage 2	5.9	–	5.7	3.2	–	–
Sport drink	2.3	–	2.6	2.0	–	–
Soy milk (low sweet)	–	–	–	3.5	–	–
Soybean milk	3.4	–	3.7	–	–	–
Orange juice	4.1	–	5.7	0.4	–	–
Grape juice	5.7	–	5.6	–	–	–
Coffee milk beverage	–	–	–	4.8	–	2.7
Brown rice liquid	–	–	–	10.0	–	–
Yogurt milk 1[b]	2.1	1.7	–	–	–	0.2
Yogurt milk 2	2.0	1.7	–	5.3	–	0.2
Fermented rice wine	27.9	–	–	–	–	–
Corn can	0.3	–	–	6.2	–	–
Black vinegar[c]	4.6	–	4.0	–	–	–
Soy sauce	5.9	–	5.7	–	–	–
Red wine 1[d]	0.1	–	–	–	–	–
Red wine 2	0.6	–	0.6	–	–	–
White wine	2.2	–	4.5	–	–	–
Japanese sake 1[e]	1.9	–	0.6	–	–	–
Japanese sake 2	1.9	–	–	–	1.3	–
Paste/Solid Foods						
Potato chips	0.5	–	–	0.7	–	–
Snack 1[f]	2.4	–	–	22.2	–	–

(Table 2) cont.....

Beverage/Food	Sugar Content in Foods (%, g/100 mL; g/100 g Sample)					
	Glc	Gal	Fru	Suc	Mal	Lac
Snack 2[g]	–	–	–	26.8	–	5.6
Snack 3[h]	7.4	–	–	29.8	–	–
Chocolate candy	–	–	–	56.9	–	8.9
Chocolate	–	–	–	36.2	–	9.8
Malt cookie	–	–	–	8.9	35.0	–
Nougat candy	–	–	–	5.8	37.5	3.7
Mashmellow	1.9	–	–	32.1	32.8	–
Herb pill[i]	8.5	–	5.9	60.1	–	–
Raisin	33.7	–	39.0	–	–	–
Dried strawberry	32.5	–	28.2	5.9	–	–
Vitamin C tablet	8.4	–	–	75.7	–	–
Preserved plum	27.6	–	23.5	–	–	–
Noodles surface sauce	–	–	–	6.2	–	–
Mushrooms seasoning	1.8	–	–	24.4	–	–
Seasoning soup block	0.3	–	–	10.3	–	–
BBQ sauce	5.3	–	7.8	31.4	–	–
Soy sauce paste 1	8.6	0.5	7.6	9.5	–	–
Soy sauce paste 2	1.3	–	2.0	2.0	–	–
Brown sugar paste	22.7	–	16.9	30.9	–	–
Honey	39.1	–	35.0	–	–	–

[a] Dong-gua tea in Chinese, [b] no added sugar, [c] made from rice, [d] low sweet, [e] made from Japan, [f] fish flavor, [g] cream coconut flavor, [h] egg flavor, [i] *Crataegus laevigata* (Hawthorn).

Some recent studies [11, 12] indicated the relationship between cancer and high sugar intake. This issue has been noticed by the government for policy implementation. The appropriate sugar intake is 25 grams per day according to the scientific recommendation by the WHO. However, there are many kinds of beverages and foods that have shown the exact amount of added sugar and other nutrition facts, but only label high, medium, and low sugar level, for example, the freshly prepared drinks in the street shops and traditional markets. Therefore, we used qNMR method to analyze some freshly prepared drinks in traditional markets.

We have measured six common sugars in freshly prepared dairy beverage. Beverage sample (50 μL) was taken and dried for NMR measurement directly. Pretreatment or dilution of the beverage sample was not required in this typical analysis. The process time was shorter than 1 h. Table **3** shows the contents of 6 common sugars in soybean milk (soya milk, Dow-jiang in Chinese). The sugar contents in the samples of soybean milk are denoted as H (high), M (medium), L (low), and F (free) in parentheses, respectively. This method provides a convenient and rapid tool for quantifying sugar ingredients in beverages and foods. In addition, the samples were also treated with NAIM labeling method (see following section), and the amounts of sugar ingredients were deduced from the ^{1}H-NMR spectral analysis. The values of sugar content are consistent between two measurements. According to our analyses, one may still intake excessive sugar, predominating in sucrose, over the daily need (25 g) as recommended by WHO and nutritionists, by drinking a cup (500 mL) of the allegedly high-sugar-content or medium-sugar-content beverage.

Table 3. Quantitative analysis of six common sugars in beverages using ^{1}H-NMR spectrometric measurement (600 MHz) in D_2O (1.0 mL) solution containing 0.03% $(CH_3)_2SO$ as internal standard.

Sugar Sample[a]	Glc	Gal	Fru	Mal	Lac	Suc	Total Sugar[b] (g /100 mL)
Soymilk (H)	< 0.1	0	0	0.1	0	15.3	15.4
Soymilk (M)	< 0.1	0	0	0.1	0	8.8	8.9
Soymilk (L)	< 0.1	0	0	0.1	0	3	3.1
Soymilk (F)	< 0.1	0	0	0.1	0	0.2	0.3

[a] The sample of 50 μL was taken for measurement, and the amounts of sugar ingredients were deduced from the integration areas in ^{1}H-NMR spectra. [b] The value of sugar content is the average of two measurements.

NMR Spectral Analysis of Mixed Sugars in Foods

The ^{1}H-NMR spectrum of a mixture of six common sugars (Glc, Gal, Fru, Mal, Lac and Suc) is shown in Fig. (**3**). In this spectrum, fructose has no distinct peak in the region of 4.13 ppm for quantification. Galactose can be quantified by integrating its anomeric protons (H-1α and H-1β). However, glucose and maltose cannot be quantified because their anomeric protons (H-1β) around 4.70 ppm overlapped. The similar phenomena of overlapping signals occur at 5.40 ppm for Suc (GlcH-1α) and Mal (GlcH-1'α), as well as around 5.25 ppm for Glc (H-1α), Lac (H-1α) and Mal (H-1α).

The ^{1}H-NMR spectra of cheese cookie (Fig. **4A**), instant coffee (Fig. **4B**), strawberry flavor jam (Fig. **4C**), maple sugar syrup (Fig. **4D**), yogurt milk (Fig. **4E**) and collagen powder (Fig. **4F**) were also recorded. Due to the overlapped signals of mixed sugars in the ^{1}H-NMR spectra of these food samples, it is difficult to differential what kind of sugar components, not to mention the accurate quantification of individual sugar. Instead of using qNMR method for direct measurement of six common sugars in foods, we pursued a more precise method assisted by NAIM derivatization. The Glc, Gal, Mal and Lac in food sample were converted to the corresponding NAIM derivatives on treatment with a NAIM labeling kit. The sugar-NAIM derivatives were readily distinguished by their characteristic signals in the ^{1}H-NMR spectrum. The preparation and calculation of each sugar-NAIM derivative are described in details in the next section.

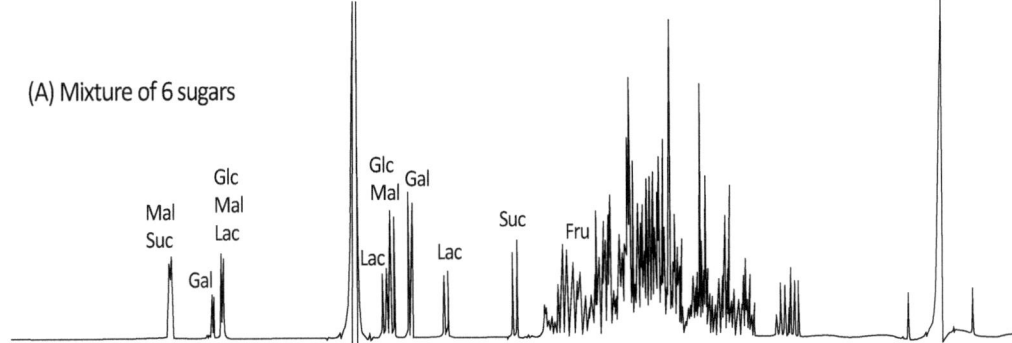

Fig. (3). A mixture of 6 sugars (Glc, Gal, Fru, Mal, Lac and Suc, 5 mg of each sugar). ^{1}H-NMR spectra (600 MHz) in D$_2$O (1.0 mL) solution containing 0.03% DMSO as internal standard. The signal of HDO was set at δ 4.80 ppm, and the signal of (CH$_3$)$_2$SO occurred at δ 2.73 ppm.

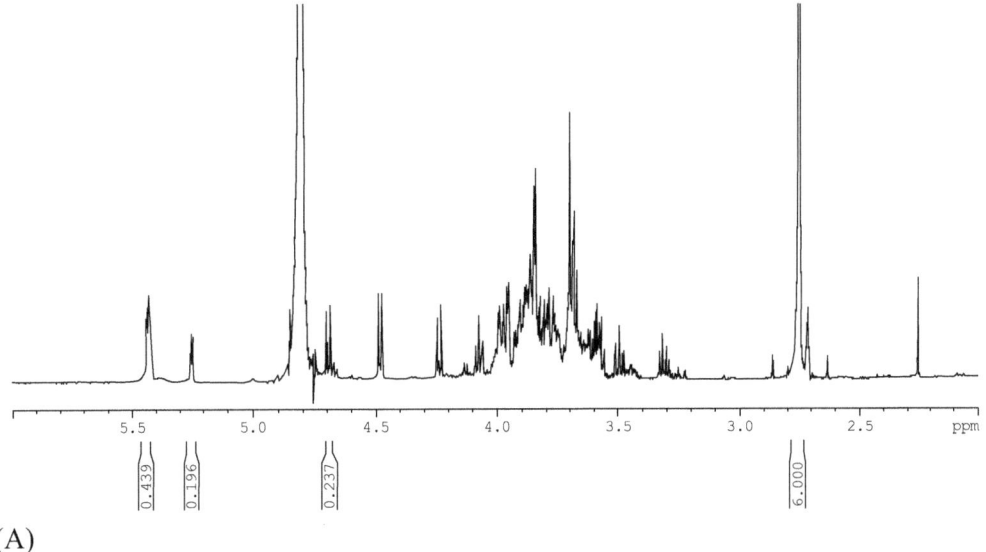

(A)

Fig. 4 cont.....

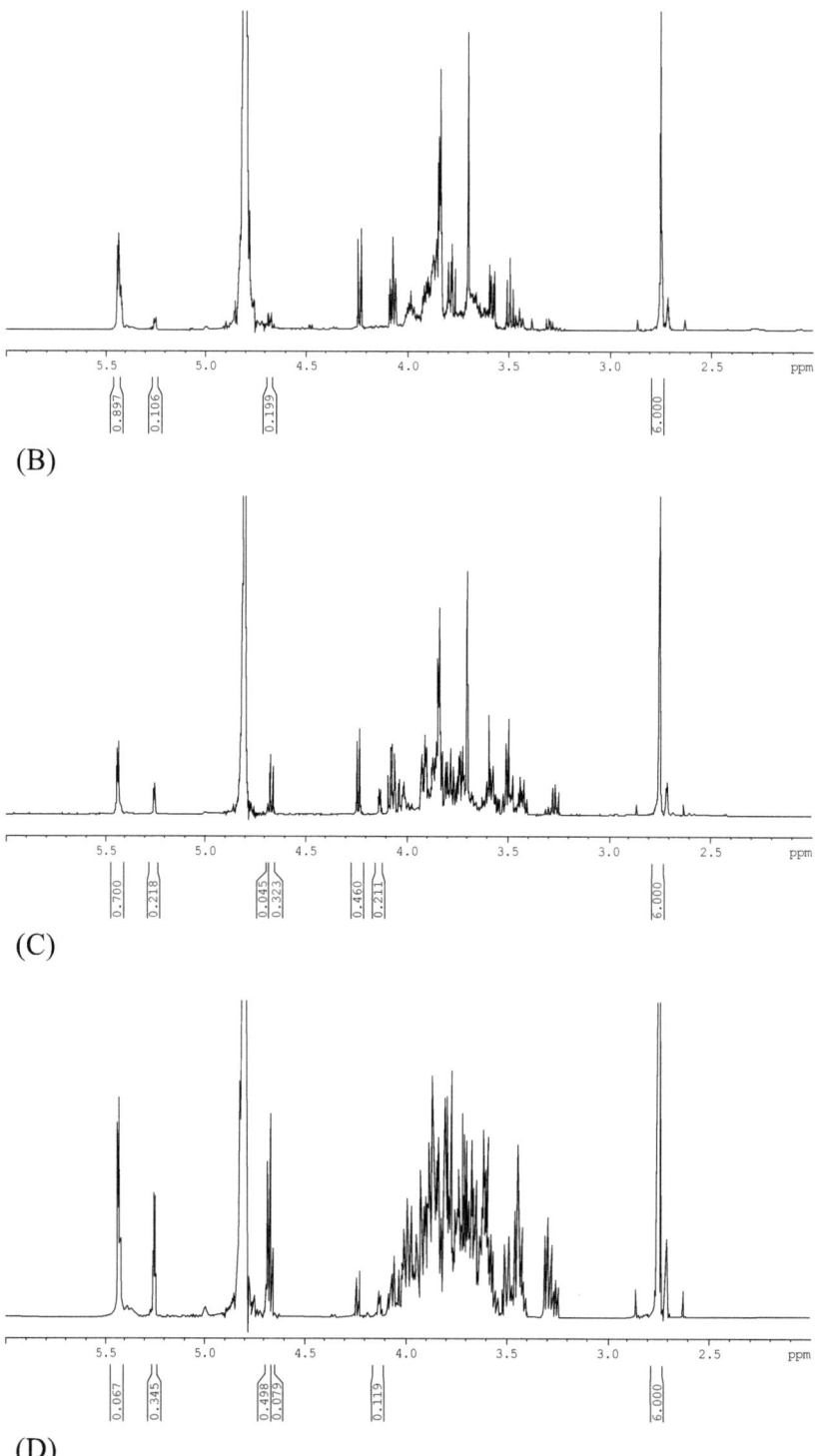

(B)

(C)

(D)

Fig. 4 cont.....

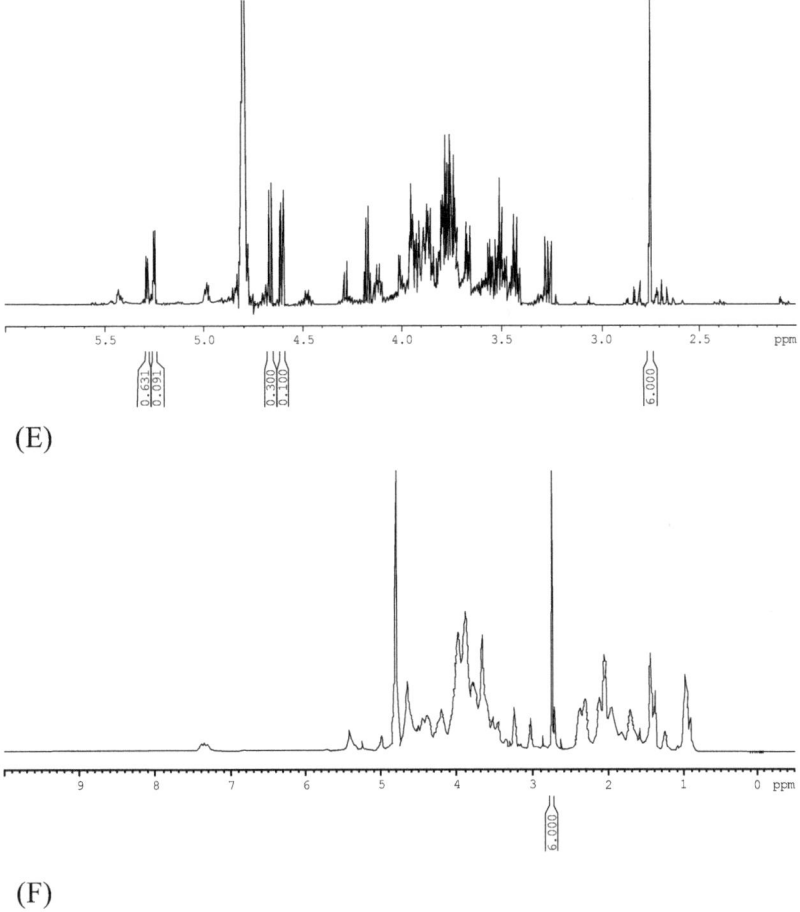

Fig. (4). ¹H-NMR spectra (600 MHz, in D$_2$O solution containing 0.03% DMSO as internal standard) of food samples: **(A)** cheese cookie containing Glc, Lac, Suc and Mal; **(B)** instant coffee (3-in-1 coffee) containing Glc, Suc, Mal and Lac (trace); **(C)** strawberry flavor jam containing Glc, Suc and Mal; **(D)** maple sugar syrup containing Glc, Suc and Mal; **(E)** yogurt milk containing Glc, Gal, Suc and Lac; **(F)** collagen powder containing small amounts of Mal and Suc.

Introduction of NAIM Labeled Sugar

Since carbohydrates lack responsive chromophores, therefore, the prior introduction of chromophore/fluorophore to carbohydrates is usually required for the detection in chromatography and electrophoresis [13]. For example, analyses of carbohydrates are often performed by labeling with appropriate reagents of 2-aminobenzamide (2-AB), 2-aminopyridine (2-AP), 1-phenyl-3-methyl-5-pyrazolone (PMP) or NADA [14] to form the derivatives that are responsive to ultraviolet-visual (UV-vis) or fluorescence detection. We have previously demonstrated that labeling aldoses with NADA *via* an iodine-promoted oxidative condensation reaction to form the NAIM derivatives is a highly sensitive method for HPLC and mass analyses [15, 16]. For example, the sugar composition in beverages and dietary foods can be determined by HPLC analysis *via* their NAIM derivatives.

General Procedure for Preparation of Sugar-NAIM Derivatives

According to the previously reported procedures [15], a mixture of glucose (2.0 mg, 11 µmol), NADA (2.0 mg, 13 µmol) and iodine (2.0 mg, 8 µmol) in AcOH (1.0 mL) was stirred at room temperature. The reaction was completed in less than 1 h as indicated by the TLC or NMR analysis. The mixture was concentrated by rotary evaporation under reduced pressure to give the sample of Glc-NAIM derivative, which was directly subjected to ^{1}H-NMR measurement without further purification. This reaction protocol is applicable to prepare other sugar-NAIM derivatives, including those of mixed sugars, in smaller quantities.

Alternatively, mono- and disaccharides were converted to the sugar-NAIM samples by using a NAIM labeling kit that consists of three vials. In brief, vial **A** containing NADA (3.0 mg) and vial **B** containing iodine (1.0 mg) in AcOH solution (1.0 mL) were used for conversion of reducing sugars to the NAIM derivatives. Vial **C** containing D_2O (1.0 mL) and a small amount (0.03−0.1%) of DMSO as the internal standard was used in recording ^{1}H-NMR spectra.

NAIM Derivatization and NMR Spectrometric Data of Aldo-Sugars

An aldose molecule exists inherently in solution as a mixture of the α and β anomeric isomers, displaying a rather complicated ^{1}H-NMR spectrum (Fig. **2**). The transformation of both aldose anomers to a single NAIM compound would simplify the ^{1}H-NMR analysis. An aldose (2.0 mg) was generally converted to the NAIM derivative at room temperature in less than 1 h by using a NAIM labeling kit that contains the reagents of NADA and iodine in acetic acid. After the reaction, acetic acid was removed by rotatory evaporation under reduced pressure, and the residue of sugar-NAIM derivative without further purification was dissolved in D_2O for recording the ^{1}H-NMR spectrum. Instead of using the conventional but less accessible reagent, trimethylsilylpropanoic acid (TMSP), the readily available and cost-effective reagent DMSO was applied as an internal standard, which showed the two methyl groups as a singlet at δ 2.73 ppm. The NAIM derivatives of several mono- and disaccharides, including Glc, Gal, mannose (Man), rhamnose (Rha), fucose (Fuc), arabinose (Ara), xylose (Xyl), glucuronic acid (GlcUA), *N*-acetylglucose (GlcNAc), Mal, and Lac were individually prepared and subjected to ^{1}H-NMR spectral analyses. Table **4** lists the characteristic proton signals of these sugar-NAIM compounds.

Table 4. ¹H-NMR data (600 MHz, D₂O) of the NAIM derivatives prepared from common mono- and disaccharides[a].

Compound	Chemical Shift (δ, ppm)				
	H-2	H-3	H-4	H-5	H-6
Glc-NAIM	5.38	4.4	3.85	3.77	3.75, 3.62
Gal-NAIM	5.60	4.15	4.06	3.95	3.80, 3.80
Man-NAIM	5.29	4.69	4.28	3.93	3.87, 3.73
Rha-NAIM	5.13	4.33	3.95	3.72	1.36
Fuc-NAIM	5.59	4.20	4.16	3.75	1.35
Ara-NAIM	5.49	4.11	3.97	3.96, 3.80	–
Rib-NAIM	5.31	4.15	3.76	3.80, 3.68	–
Xyl NAIM	5.43	4.21	3.99	3.79, 3.72	–
GlcUA-NAIM	5.53	4.41	4.13	4.31	–
GlcNAc-NAIM	5.44	4.51	3.9	3.84	3.73, 3.60
Mal-NAIM	5.52	3.89	3.66	4.11	3.93, 3.82
Lac-NAIM	5.56	4.43	4.14	4.08	3.93, 3.84

[a] The signal of HDO was set at δ 4.80 ppm, and the internal standard (CH₃)₂SO (0.03%, v/v) occurred at δ 2.73 ppm.

In above section, we demonstrate that qNMR spectroscopy is a powerful method for simultaneous identification and quantification of six common sugars in foods. One-dimensional (1D) ¹H-NMR spectra are also used in the routine quantitative analysis of sugars due to a short acquisition time. However, the simultaneous quantification of all sugar ingredients from a single 1D ¹H-NMR spectrum is still difficult due to overlap of the proton signals. Alternatively, we present in this chapter a convenient method for the routine quantification of six common sugars in beverages and foods *via* NAIM derivatization for 1D ¹H-NMR spectrometric analysis (see following section and workflow 2).

The ¹H-NMR spectrum of Glc-NAIM (Fig. **5A**) was significantly simplified, and the H-2 shifted downfield to δ 5.38 as a doublet ($J = 5.4$ Hz). The characteristic H-2 of Gal-NAIM (Fig. **5B**) appeared at δ 5.54 (d, $J = 1.8$ Hz). The disaccharide derivative Mal-NAIM (Fig. **5C**) exhibited H-2 at δ 5.52 (d, $J = 1.8$ Hz) and the glycosidic proton (H-1′) at δ 5.25 (d, $J = 3.6$ Hz), whereas Lac-NAIM (Fig. **5D**) displayed H-2 at δ 5.56 (d, $J = 4.2$ Hz) and H-1′ at δ 4.59 (d, $J = 7.8$ Hz). These sugar-NAIM derivatives consistently showed the characteristic patterns of C-2 protons in the region of 5.1–5.7 ppm in addition to other well recognizable proton signals in the ¹H-NMR spectra. Thus, the parental sugars could be easily inferred from their corresponding NAIM derivatives using ¹H-NMR spectrometry.

Specifically, this ¹H-NMR method is versatile to distinguish glucose from mannose (C2-epimer) and galactose (C4-epimer). Maltose and lactose were also readily differentiated by the ¹H-NMR spectra of their NAIM derivatives.

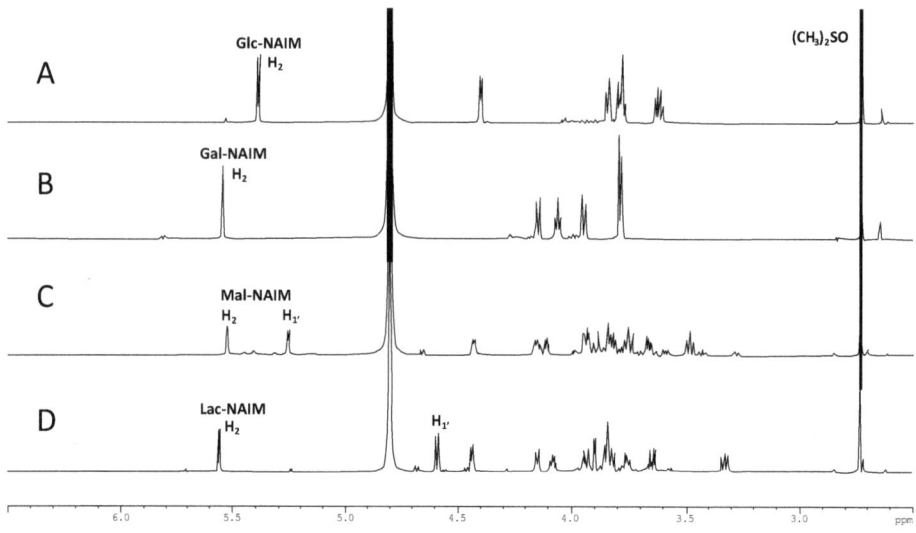

Fig. (5). ¹H-NMR spectra (600 MHz) in D_2O solution (1.0 mL) containing 0.03% DMSO as internal standard: (A) Glc-NAIM, (B) Gal-NAIM, (C) Mal-NAIM, and (D) Lac-NAIM. The aromatic protons of NAIM derivatives in the range of δ 7.2–8.2 ppm are not shown for clearance. The signal of HDO was set at δ 4.80 ppm, and the signal of $(CH_3)_2SO$ occurred at δ 2.73 ppm.

To demonstrate the advantage of using NAIM derivatives in ¹H-NMR analysis of sugar mixture, a sample containing 4 aldoses (Glc, Gal, Mal, and Lac) in equal amounts was subjected to NAIM derivatization using a NAIM labeling kit, followed by quantitative analysis using ¹H-NMR spectrometry. The parental sugars Glc, Mal, and Lac could not be easily quantified because their C-1 protons overlapped on the same position (Fig. 3). In contrast, the sugar-NAIM derivatives were easily distinguished by their C-2 protons with the diagnostic patterns at distinct chemical shifts. Taking the integration area of the $(CH_3)_2SO$ peak from δ 2.79 to 2.73 ppm for the two methyl groups as a reference, one could calculate the amount of each sugar-NAIM derivative from its H-2 signal. In addition, the glycoside protons (H-1′) of Mal-NAIM (at 5.25 ppm) and Lac-NAIM (at 4.59 ppm) could also be used for quantitative analysis.

NMR Spectrometric Analysis of Mixed Sugars *via* NAIM Derivatization

We examined the ¹H-NMR spectrum of a mixture containing six common sugars (Glc, Gal, Fru, Mal, Lac, and Suc). In this spectrum (Fig. 6), the sugar-NAIM derivatives were readily distinguished by their characteristic signals in the ¹H-NMR spectrum. Taking the integration areas of the characteristic proton signals, one can calculate the amount of each sugar-NAIM derivative, for example, from the H-2 signals of Glc-NAIM at δ 5.47, Gal-NAIM at δ 5.65, Mal-NAIM at δ 5.57, and Lac-NAIM at δ 5.59 ppm. The signals at 5.25 and 4.59 ppm were also diagnostic for H-1' of Mal-NAIM and Lac-NAIM, respectively. It was noted that fructose has no distinct peak at 5.1~5.7 ppm for identification; however, in an acidic condition on treatment with NAIM kit, a partial conversion of Fru to Fru-enamine (at 5.27 ppm) happened [17], so that we added this signal for Fru identification and quantification. One of the proton signals of unmodified Fru remains at 4.13 ppm. Sucrose, a non-reducing sugar, was retained without oxidative condensation by NADA under such reaction conditions. Nonetheless, sucrose can be identified by its glycosidic proton (H-1) at 5.44 ppm and another proton at 4.23 ppm.

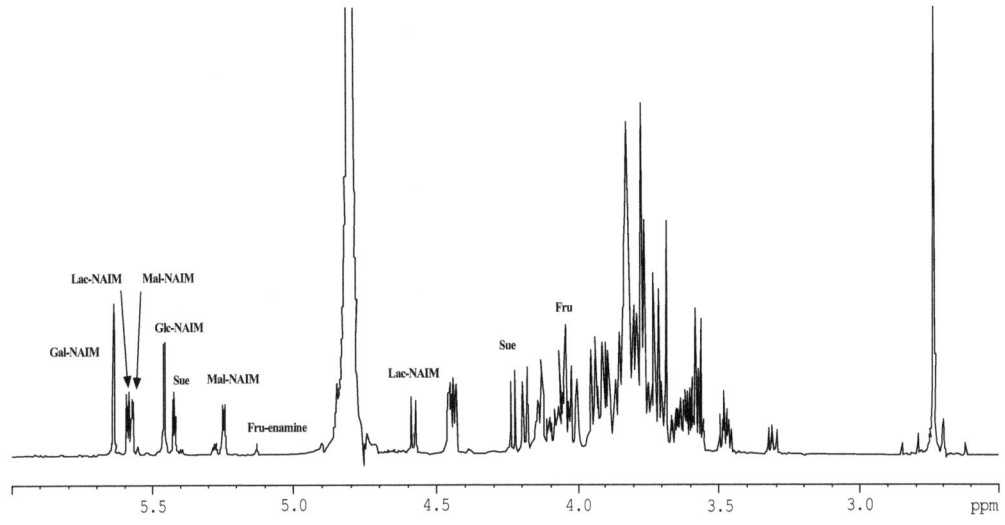

Fig. (6). A mixture of 6 sugars (Glc, Gal, Fru, Mal, Lac and Suc, 5 mg of each sugar) was treated with NAIM kit. This spectrum shows that Glc, Gal, Mal and Lac were converted into Glc-NAIM, Gal-NAIM, Mal-NAIM, Lac-NAIM, Fru was partially converted to Fru-enamine (at 5.27 ppm), but unmodified Suc remained. The aromatic protons of the NAIM derivatives in the range of δ 7.2–8.2 ppm are not shown for clearance. ¹H-NMR spectra (600 MHz) in D_2O (1.0 mL) solution containing 0.03% DMSO as internal standard. The signal of HDO was set at δ 4.80 ppm, and the signal of $(CH_3)_2SO$ occurred at δ 2.73 ppm.

Workflow 2: Measurement NAIM Labeled Sugars, Fru and Suc in Foods

In this section, ¹H-NMR determination of six common sugars (Glc, Gal, Fru, Suc, Mal, Lac) in foods was based on the NAIM derived sugars (Glc-NAIM, Gal-NAIM, Mal-NAIM, Lac-NAIM), Fru and Suc, The flowchart (Fig. 7) shows the steps of sample preparation, NMR processing and statistical analysis.

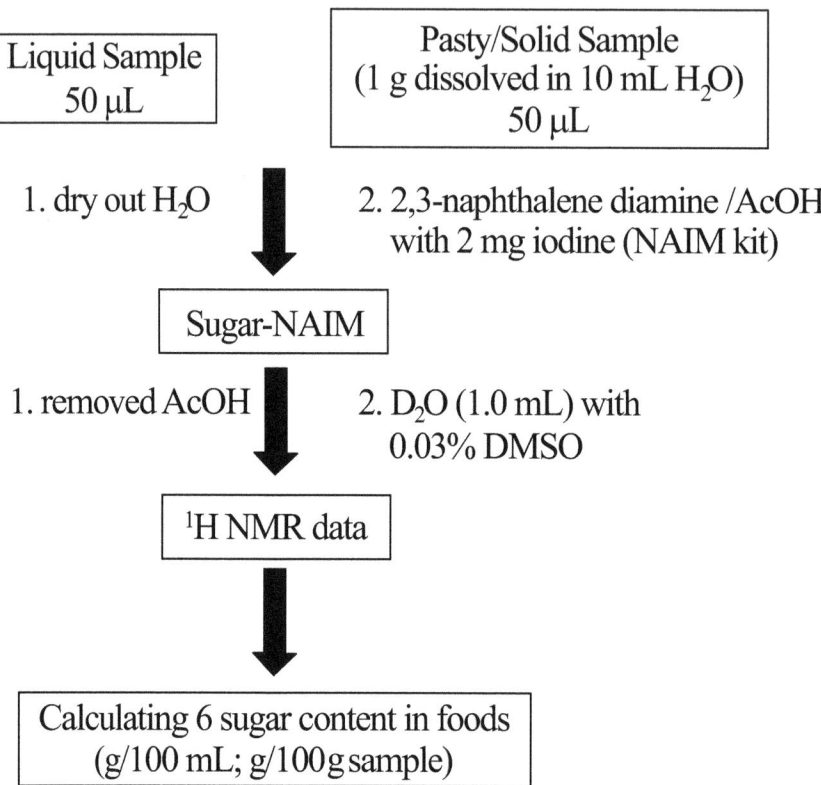

Fig. (7). Workflow of using ¹H-NMR for determination of 4 NAIM derived sugars, Fru and Suc.

Sample Preparation

The materials for sugars, iodine, AcOH, NADA, and deuterium solvents were purchased from Merck & Co., Inc. (Darmstadt, Germany). Beverage and food samples were purchased from stores in Taipei City, Taiwan. The NAIM labeling kit used in this study was a gift from Sugarlighter Co., Inc. (New Taipei City, Taiwan). This kit is suitable for analysis of sugar ingredients for 3 mg of the total amount in a sample. Four standard sugars (Glc, Gal, Mal, and Lac) were prepared in varied concentrations for quantification (5.0, 2.5, 1.25 and 0.25 mg, respectively), Beverage and food samples were taken 50 μL and the solvent was removed. For instance, solution-state sample was taken 50 μL directly, and 1.0 gram of solid-state/paste sample was dissolved in 10.0 mL of H_2O, and the solution (50 μL) was taken to remove solvent for further preparation. The detailed procedure for preparation of sugar-NAIM derivatives is shown in the above section.

NMR Experimental Process

The solvent and internal standard for NMR experiment were same as that mentioned above. The above-prepared NAIM-derivatized sample was dissolved and put into a 5-mm NMR tube for ^{1}H-NMR spectral measurement. The ^{1}H-NMR spectra were recorded on a Bruker AV600 MHz NMR spectrometer. Quantification of sugars was based on the integration areas of the characteristic proton signals (*e.g.*, H-2 in Glc-NAIM) by comparison with that of DMSO (integration region from δ 2.79 to 2.73 ppm for six protons of the two methyl groups). ^{1}H-NMR acquisition parameters were the same as that mentioned above.

Statistical Analysis

In general, calculation of 6 common sugars in foods is based on ^{1}H-NMR method. HDO signal is fixed at 4.80 ppm and DMSO (δ 2.79 to 2.73 ppm for six protons of the two methyl groups). The integration area was set at 6.00 based on DMSO protons. The corresponding calculation peaks of 6 common sugars are shown in Table 5. We also use a coefficient K to facilitate the calculation of 6 common sugars in foods. In briefly, using qNMR concept the molar proportion of integration region between DMSO (0.03% in 1.0 mL of D_2O) and sugars (Glc-NAIM and Gal-NAIM have molecular weight 318 g/mol; Fru has molecular weight 180 g/mol; Mal-NAIM, Lac-NAIM have molecular weight 480 g/mol, Suc has molecular weight 342 g/mol, respectively) was calculated to converse to a unit of g sugar/100 mL or g sugar/100 g in food. To simplify the calculation of six common sugars in foods, we provided a formula as following to help the sugars quantification in foods. The content of six common sugars in foods (in gram/100 mL or gram/100 g) = K (sugar coefficient) * IA (integration area of sugar in ^{1}H NMR) * D (times of dilution from original sample/food, among of solution-state sample = 1.000; solid-state or pasty samples = 10.000). The empirical equation (coefficient K) is to facilitate quantification of each common sugar by the integration area of the selected proton signals (Table 1 and Table 5).

Table 5. The ^{1}H-NMR integration regions and coefficient K of 6 common sugars, in which Glc, Gal, Mal and Lac are converted to the NAIM derivatives.

Sugar Type	Integral Regions (δ, ppm)	Coefficient (K)
Glc-NAIM	5.53	1.3
Gal-NAIM	5.72	1.3
Fru	4.13 + 5.27	1.5
Suc	5.44	1.5
Mal-NAIM	5.25	2.5
Lac-NAIM	5.67	2.5

We have measured six common sugars in dairy beverages and foods using the NAIM method (Table 6). Beverage sample (50 µL) was taken and directly treated with a NAIM labeling kit. Solid-state/paste sample 50 µL (from 10 mL of dilution solution) was taken and treated with a NAIM labeling kit. The components of reducing sugars in beverages and foods were converted to the corresponding NAIM derivatives at room temperature in 1 h. The resulting solution was concentrated under reduced pressure by rotary evaporation, and the residue was dissolved in deuteriated water (D_2O, 1.0 mL with 0.03% of DMSO) for subsequent measurement. We used qNMR method and statistical analysis to calculate 6 common sugars in foods. Taking the integration areas of the characteristic proton signals, one can calculate the amount of each sugar. For example, in the solution-state samples: orange juice has 10.2 g sugar/100 mL comprising Glc (4.1 g), Fru (5.7 g) and Suc (0.4 g); yogurt milk 2 has 9.2 g sugar/100 mL comprising Glc (2.0 g), Gal (1.7 g), Suc (5.3 g) and Lac (0.2 g); refreshment beverage 1 has 16.8 g sugar/100 mL comprising Glc (2.8 g), Fru (3.1 g) and Suc (10.9 g); sport drink has 6.9 g sugar/100 mL comprising Glc (2.3 g), Fru (2.6 g) and Suc (2.0 g). In solid-state/paste food samples: cookie (squid flavor) has 14.6 g sugar/100 g comprising Glc (0.1 g), Suc (11.7 g) and Mal (2.8 g), potato chip has 1.0 g sugar/100 g comprising Glc (0.5 g) and Suc (0.5 g), chocolate bar has 46.0 g sugar/100 g comprising Suc (36.2 g) and Lac (9.8 g), jam (strawberry fruit) has 51.9 g sugar/100 g comprising Glc (21.3 g), Fru (11.8 g) and Suc (18.8 g), syrup (for muffin seasoning) has 66.5 g sugar/100 g comprising Glc (9.5 g), Fru (5.8 g), Suc (5.3) and Mal (45.9 g), soy sauce paste 2 has 5.3 g sugar/100 g comprising Glc (1.3 g), Fru (2.0 g) and Suc (2.0 g), honey has 77.1 g sugar/100 g comprising Glc (41.1 g) and Fru (36.0 g), respectively. The results are consistent with that shown in Table 2 (native/no NAIM derived sugars in beverages and foods by qNMR); nonetheless, the NAIM method has advantage in resolving the overlapped NMR signals of native sugars in beverages and foods.

Table 6. Using NAIM derivatization method to measure six common sugars in dairy beverages and foods by qNMR.

Beverage/Food	Sugar Content in Foods (%, g/100 mL; g/100 g Sample)					
	Glc	Gal	Fru	Suc	Mal	Lac
Tung-Gua drink[a]	0.1	–	–	7.4	–	–
Lemon beverage	0.3	–	0.2	5.0	–	–
Lemon black tea	3.5	–	3.3	1.1	–	–
Honey-plum vinegar	2.6	–	2.1	5.7	–	–
Refreshment beverage 1	3.5	–	2.4	10.9	–	–
Refreshment beverage 2	6.7	–	4.8	3.2	–	–

(Table 6) cont.....

Beverage/Food	Sugar Content in Foods (%, g/100 mL; g/100 g Sample)					
	Glc	Gal	Fru	Suc	Mal	Lac
Sport drink	2.8	–	2.7	1.8	–	–
Soy milk (low sweet)	–	–	–	3.5	–	–
Soybean milk	3.5	–	3.6	–	–	–
Orange juice	4.1	–	5.7	0.4	–	–
Grape juice	5.8	–	5.6	–	–	–
Coffee milk beverage	–	–	–	4.8	–	2.7
Brown rice liquid	–	–	–	10.2	–	–
Yogurt milk 1[b]	2.1	1.7	–	–	–	0.2
Yogurt milk 2	2.0	1.7	–	5.3	–	0.2
Fermented rice wine	28.2	–	–	–	–	–
Corn can	0.3	–	–	6.0	–	–
Black vinegar[c]	4.6	–	4.0	–	–	–
Soy sauce	5.9	–	5.7	–	–	–
Red wine[d]	0.1	–	–	–	–	–
Red wine	0.6	–	0.6	–	–	–
White wine	2.2	–	4.5	–	–	–
Japanese sake 1[e]	1.9	–	0.6	–	–	–
Japanese sake 2	1.9	–	–	–	1.3	–
Paste/Solid Foods	-	-	-	-	-	-
Potato chips	0.5	–	–	0.5	–	–
Snack 1[f]	2.4	–	–	22.2	–	–
Snack 2[g]	–	–	–	27.0	–	5.6
Snack 3[h]	7.4	–	–	29.8	–	–
Chocolate candy	–	–	–	56.9	–	8.9
Chocolate	–	–	–	36.2	–	9.8
Malt cookie	–	–	–	8.9	35.1	–
Cookie[i]	0.1	–	–	11.7	–	2.8
Nougat candy	–	–	–	5.8	37.5	3.7
Mashmellow	2.0	–	–	32.1	32.8	–
Herb pill[j]	8.5	–	5.9	60.0	–	–
Raisin	33.7	–	39.5	–	–	–
Dried strawberry	32.5	–	28.2	5.9	–	–
Vitamin C tablet	8.4	–	–	75.0	–	–

(Table 6) cont.....

Beverage/Food	Sugar Content in Foods (%, g/100 mL; g/100 g Sample)					
	Glc	Gal	Fru	Suc	Mal	Lac
Preserved plum	27.6	–	23.5	–	–	–
Noodles surface sauce	–	–	–	6.2	–	–
Mushrooms seasoning	1.8	–	–	24.4	–	–
Seasoning soup block	0.3	–	–	10.3	–	–
BBQ sauce	5.3	–	7.8	31.4	–	–
Soy sauce paste 1	8.6	0.5	7.6	9.5	–	–
Soy sauce paste 2	1.3	–	2.0	2.0	–	–
Brown sugar paste	22.7	–	16.9	30.9	–	–
Strawberry jam	21.3	–	11.8	18.8	–	–
Muffin syrup	9.5	–	5.8	5.3	–	45.9
Honey	41.1	–	36.0	–	–	–

[a] Dong-gua tea in Chinese, [b] no added sugar, [c] made from rice, [d] low sweet, [e] made from Japan, [f] fish flavor, [g] cream coconut flavor, [h] egg flavor, [i] squid flavor, [j] Crataegus laevigata.

DISCUSSION

The qNMR technique has been designed for direct quantification of mixture components without the requirement of any sample preparation steps [18]. In comparison, pretreatment of sugars to the NAIM derivatives can resolve the overlaps of proton signals to simplify the quantitative analysis by NMR technique. Our present methods (NAIM labeled sugars and non-labeled/native sugars) are suitable to quantify individual sugar ingredients, which would supposedly be conducted in the presence of other non-sugar components in the beverage and food samples.

We have also previously performed HPLC method to quantify the six common sugars in beverages [19]. The NAIM derivatives of Glc, Gal, Lac, and Mal can be quantified by using UV or fluorescence detector, though a less sensitive refractive index (RI) detector is needed to quantify the unmodified Fru and Suc that have no chromophore. The HPLC analysis usually takes a longer elution time than recording an ^{1}H-NMR spectrum, and the HPLC method also requires parental sugars and sugar-NAIM standards for calibration in every quantitative analysis.

CONCLUSIONS

Here, we demonstrate an efficient ^{1}H-NMR spectrometric method for routine quantification of six common sugars in beverages and foods. In this typical analysis, pretreatment or dilution of the beverage sample is not required. The beverage sample (50 µL), *i.e.* beverage in this study was simply treated with a NAIM labeling kit containing the reagents of NADA and iodine in acetic acid to convert the four reducing sugars (*i.e.* glucose, galactose, lactose and maltose) to their corresponding sugar-NAIM derivatives, along with unchanged sucrose and partially modified fructose. After rotary evaporation under reduced pressure, a 1D ^{1}H-NMR spectrum of the residual sample, without further purification, was sufficient for quantitative determination of the six common sugars because the sugar-NAIM compounds exhibited the diagnostic H-2 protons in much better resolution than their parental sugars, which are complicated due to having both α and β anomers. In this particular case, we further established the calibration lines [17] and empirical equations (Statistical Analysis) for facile quantification of each common sugar by the integration area of the selected proton signals (Table **1** and Table **5**) relative to that of DMSO (0.03% in D_2O solution). DMSO appears to be a suitable internal standard that is more cost-effective than TMSP. In this study, we also demonstrate that the monosaccharides can be easily identified and quantified by the NAIM derivatization for ^{1}H-NMR spectral analysis (Table **3**).

To cope with the proposal of the TFDA on listing the amounts of six common sugars on "Nutrition Facts Panel" of beverages and foods, the non-labeled qNMR method takes a short process time (< 1 h), but the proton signals in some samples might overlap. Although the NAIM method takes a somewhat longer process time, the NAIM reaction is smoothly performed, and the product without further purification is directly subjected to the ^{1}H-NMR analysis after removal of the AcOH solvent. This operation renders the anomeric isomers of an aldose to a single NAIM derivative that shows the characteristic H-2 signal at downfield for diagnosis and quantitative analysis by ^{1}H-NMR spectrometry. Our results suggest that a simple treatment of beverage and food with the NAIM labeling method provides a more extensive success rate for the quantification of sugar ingredients. Consequently, the method combining NAIM derivatization and ^{1}H-NMR analysis is also potentially useful for the fingerprinting/profiling the sugars in foods.

ABBREVIATIONS

2-AB	2-aminobenzamide;
AcOH	acetic acid; 2-AP, 2-aminopyridine; Ara, arabinose;
dd-H_2O	double-distilled water; D_2O, deuterium water;
DMSO	dimethylsulfoxide; Fru, fructose; Fuc, fucose; Gal, galactose; GI, glycemic index; Glc, glucose; GlcUA, glucuronic acid;
GlcNAc	*N*-acetylglucosamine;
HFCS	high fructose corn syrup;
HPAE−PAD	high-performance anion-exchange chromatography with pulsed amperometric detection; HPLC, high performance liquid chromatography; Lac, lactose; Man, mannose; Mal, maltose;

NADA	2,3-naphthalene diamine;
NAIM	naphthimidazole;
NMR	nuclear magnetic resonance;
PMP	1-phenyl-3-methyl-5-pyrazolone;
qNMR	quantitative NMR; Rha, rhamnose;
RI	refractive index; Suc, sucrose;
TFDA	Taiwan Food & Drug Administration; TLC, thin-layer chromatography; TMSP, trimethylsilylpropanoic acid; UV-vis, ultraviolet-visible; WHO, World Health Organization; Xyl, xylose.

CONSENT FOR PUBLICATION

Not applicable.

CONFLICT OF INTEREST

The authors confirm that this chapter contents have no conflict of interest.

ACKNOWLEDGEMENTS

Y.-T. Chen and S.-H. Wang contribute to execution of experiments, data acquisition and statistical analysis. W.-B. Yang provides administrative, technical and writing manuscript. The authors thank Dr. Chi-Fon Chang for the technical assistance on NMR core facility in Genomics Research Center, and Sugarlighter Co., Inc. (New Taipei City, Taiwan) for providing us the NAIM labeling kit. This work was supported by Genomics Research Center, Academia Sinica.

REFERENCES

[1] Montero, C.M.; Dodero, M.C.R.; Sánchez, D.A.G.; Barroso, C.G. Analysis of low molecular weight carbohydrates in food and beverages: a review. *Chromatographia*, **2004**, *59*, 15-30.

[2] Soga, T. Analysis of carbohydrates in food and beverages by HPLC and CE. *J. Chromatogr. Libr.*, **2002**, *66*, 483-502.
[http://dx.doi.org/10.1016/S0301-4770(02)80039-2]

[3] Goncalves, M.D.; Lu, C.; Tutnauer, J.; Hartman, T.E.; Hwang, S.K.; Murphy, C.J.; Pauli, C.; Morris, R.; Taylor, S.; Bosch, K.; Yang, S.; Wang, Y.; Van Riper, J.; Lekaye, H.C.; Roper, J.; Kim, Y.; Chen, Q.; Gross, S.S.; Rhee, K.Y.; Cantley, L.C.; Yun, J. High-fructose corn syrup enhances intestinal tumor growth in mice. *Science*, **2019**, *363*(6433), 1345-1349.
[http://dx.doi.org/10.1126/science.aat8515] [PMID: 30898933]

[4] Rippe, J.M.; Angelopoulos, T.J. Relationship between added sugars consumption and chronic disease risk factors: current understanding. *Nutrients*, **2016**, *8*(11), 697-715.
[http://dx.doi.org/10.3390/nu8110697] [PMID: 27827899]

[5] Mannucci, E.; Dicembrini, I.; Lauria, A.; Pozzilli, P. Is glucose control important for prevention of cardiovascular disease in diabetes? *Diabetes Care,* **2013**, *36* Suppl. 2, S259-S263.
[http://dx.doi.org/10.2337/dcS13-2018] [PMID: 23882055]

[6] Falbe, J.; Rojas, N.; Grummon, A.H.; Madsen, K.A. Higher retail prices of sugar-sweetened beverages 3 months after implementation of an excise tax in Berkeley, California. *Am. J. Public Health,* **2015**, *105*(11), 2194-2201.
[http://dx.doi.org/10.2105/AJPH.2015.302881] [PMID: 26444622]

[7] Nishida, C.; Uauy, R.; Kumanyika, S.; Shetty, P. The joint WHO/FAO expert consultation on diet, nutrition and the prevention of chronic diseases: process, product and policy implications. *Public Health Nutr.,* **2004**, *7*(1A), 245-250.
[http://dx.doi.org/10.1079/PHN2003592] [PMID: 14972063]

[8] Weber, M.; Hellriegel, C.; Rueck, A.; Wuethrich, J.; Jenks, P. Using high-performance ^{1}H NMR (HP-qNMR®) for the certification of organic reference materials under accreditation guidelines--describing the overall process with focus on homogeneity and stability assessment. *J. Pharm. Biomed. Anal.,* **2014**, *93*, 102-110.
[http://dx.doi.org/10.1016/j.jpba.2013.09.007] [PMID: 24182847]

[9] Cao, R.; Komura, F.; Nonaka, A.; Kato, T.; Fukumashi, J.; Matsui, T. Quantitative analysis of D-(+-)-glucose in fruit juices using diffusion ordered-^{1}H nuclear magnetic resonance spectroscopy. *Anal. Sci.,* **2014**, *30*(3), 383-388.
[http://dx.doi.org/10.2116/analsci.30.383] [PMID: 24614734]

[10] Rundlöf, T.; Mathiasson, M.; Bekiroglu, S.; Hakkarainen, B.; Bowden, T.; Arvidsson, T. Survey and qualification of internal standards for quantification by ^{1}H NMR spectroscopy. *J. Pharm. Biomed. Anal.,* **2010**, *52*(5), 645-651.
[http://dx.doi.org/10.1016/j.jpba.2010.02.007] [PMID: 20207092]

[11] Hu, C.M.; Tien, S.C.; Hsieh, P.K.; Jeng, Y.M.; Chang, M.C.; Chang, Y.T.; Chen, Y.J.; Chen, Y.J.; Lee, E.Y.P.; Lee, W.H. High glucose triggers nucleotide imbalance through *O*-GlcNAcylation of key enzymes and induces KRAS mutation in pancreatic cells. *Cell Metab.,* **2019**, *29*(6), 1334-1349.e10.
[http://dx.doi.org/10.1016/j.cmet.2019.02.005] [PMID: 30853214]

[12] Goncalves, M.D.; Lu, C.; Tutnauer, J.; Hartman, T.E.; Hwang, S-K.; Murphy, C.J.; Pauli, C.; Morris, R.; Taylor, S.; Bosch, K.; Yang, S.; Wang, Y.; Van Riper, J.; Lekaye, H.C.; Roper, J.; Kim, Y.; Chen, Q.; Gross, S.S.; Rhee, K.Y.; Cantley, L.C.; Yun, J. High-fructose corn syrup enhances intestinal tumor growth in mice. *Science,* **2019**, *363*(6433), 1345-1349.
[http://dx.doi.org/10.1126/science.aat8515] [PMID: 30898933]

[13] Ruhaak, L.R.; Zauner, G.; Huhn, C.; Bruggink, C.; Deelder, A.M.; Wuhrer, M. Glycan labeling strategies and their use in identification and quantification. *Anal. Bioanal. Chem.,* **2010**, *397*(8), 3457-3481.
[http://dx.doi.org/10.1007/s00216-010-3532-z] [PMID: 20225063]

[14] Harvey, D.J. Derivatization of carbohydrates for analysis by chromatography; electrophoresis and mass spectrometry. *J. Chromatogr. B Analyt. Technol. Biomed. Life Sci.,* **2011**, *879*(17-18), 1196-1225.
[http://dx.doi.org/10.1016/j.jchromb.2010.11.010] [PMID: 21145794]

[15] Lin, C.; Lai, P.T.; Liao, S.K.; Hung, W.T.; Yang, W.B.; Fang, J.M. Using molecular iodine in direct oxidative condensation of aldoses with diamines: an improved synthesis of aldo-benzimidazoles and aldo-naphthimidazoles for carbohydrate analysis. *J. Org. Chem.,* **2008**, *73*(10), 3848-3853.
[http://dx.doi.org/10.1021/jo800234x] [PMID: 18422361]

[16] Lin, C.; Hung, W.T.; Kuo, C.Y.; Liao, K.S.; Liu, Y.C.; Yang, W.B. I_2-catalyzed oxidative condensation of aldoses with diamines: synthesis of aldo-naphthimidazoles for carbohydrate analysis. *Molecules,* **2010**, *15*(3), 1340-1353.
[http://dx.doi.org/10.3390/molecules15031340] [PMID: 20335985]

[17] Chen, Y.T.; Wang, S.H.; Hung, W.T.; Fang, J.M.; Yang, W.B. Quantitative analysis of sugar ingredients in beverages and food crops by an effective method combining naphthimidazole derivatization and ^{1}H-NMR spectrometry. *Funct. Food Health Dis.,* **2017**, *7*, 494-510.

[http://dx.doi.org/10.31989/ffhd.v7i7.348]

[18] Pauli, G.F.; Gödecke, T.; Jaki, B.U.; Lankin, D.C. Quantitative ^{1}H NMR. Development and potential of an analytical method: an update. *J. Nat. Prod.,* **2012**, *75*(4), 834-851.
[http://dx.doi.org/10.1021/np200993k] [PMID: 22482996]

[19] Hung, W.T.; Chen, Y.T.; Wang, S.H.; Liu, Y.C.; Yang, W.B. A new method for aldo-sugar analysis in beverages and dietary foods. *Funct. Food Health Dis.,* **2016**, *6*, 234-245.
[http://dx.doi.org/10.31989/ffhd.v6i4.251]

CHAPTER 2

Correlation Between VIP Scores and ¹H NMR to Extract Information of Psychological Attention Tests Applied Before and After Coffee Intake

Michel Rocha Baqueta[1], Aline Coqueiro[2], Letícia de Sousa Frutuozo[1], Paulo Henrique Março[1], Frank Duarte[3], Manuela Mandrone[4], Ferruccio Poli[4] and Patrícia Valderrama[1,*]

[1] Universidade Tecnológica Federal do Paraná, Campo Mourão, Paraná, Brazil

[2] Universidade Tecnológica Federal do Paraná, Ponta Grossa, Paraná, Brazil

[3] Faculdade União de Campo Mourão, Campo Mourão, Paraná, Brazil

[4] Università di Bologna, Bologna, Italy

Abstract: This chapter presents the correlation between coffee compounds identified by ¹H Nuclear Magnetic Resonance (¹H NMR) spectroscopy with psychological attention tests in order to verify which compounds are related to the focus and/or diffuse attention. Psychometric tests applied by a clinical psychologist, before and after coffee intake, were the focus attention AC-vector and the diffuse attention TADIM, and the focus attention TACOM-B and the diffuse attention TEDIF, respectively. Different tests to measure the attention before and after coffee intake were used to avoid learning effects. After AC-vector and TADIM tests, each volunteer consumed a total of 40 mL of coffee with different cup qualities (four different coffee blends – 10 mL per beverage) and indicated the order of preference in relation to the smell. This approach was used to create a greater metabolic variation between the samples tested, allowing to build a robust chemometric model. For each preferred coffee, a ¹H NMR spectrum was obtained and a chemometric data treatment based on Partial Least Squares (PLS) regression and Variable Importance in Projection scores (VIP scores) was used to correlate the spectra with the psychological test results and to verify which metabolites of the coffee beverage could be related to the focus or diffuse attention. In general, our results showed that coffee intake attenuated diffuse attention and improved focus attention in most volunteers. The major metabolites that contributed to both diffuse and focus attention were caffeine, trigonelline, chlorogenic acids, acetate, lipids, lactate, γ-quinide, and polysaccharides. Among metabolites exclusively important to focus attention, formate, choline, myo-inositol, citrate, and malate were the most important. Therefore, the ¹H NMR profile, in combination with chemometric tools, is interesting to assess the correlation between coffee compounds and human attention.

---

* **Corresponding author Patrícia Valderrama:** Universidade Tecnológica Federal do Paraná, Campo Mourão, Paraná, Brazil; Tel:(+55) 44 3518-1525; E-mail: pativalderrama@gmail.com

Keywords: Attention Performance, Chemometrics, Coffee, Coffee compounds, Coffee ingestion, Coffee smell, Cup quality, Metabolomics, NMR spectroscopy, Pilot Study, PLS Regression, Psychometric Tests, VIP scores.

INTRODUCTION

Coffee is globally one of the most widely-consumed beverages that contain over 1000 compounds responsible for its pleasant flavor and aroma [1]. Coffee is known for its stimulant, beneficial and nutritional properties leading researchers to seek to understand how these properties are correlated to their chemical composition. For example, coffee drinking habits have been associated with a decrease in the risk of developing Alzheimer's and Parkinson's disease [2], and a decrease in the incidence of cardiovascular disease and type 2 diabetes mellitus [3 - 5]. Recently, it was shown that the risk of Alzheimer's disease was lower in those who regularly consume coffee than those who do not drink it [6].

The bioactivities related to coffee consumption are related to several compounds. Of these, caffeine (1,3,7-trimethylxanthine) is the most widely studied. Caffeine is a psychoactive and neurostimulator substance that exerts most of its biological effects as an adenosine receptor antagonist, inducing a generally stimulating effect in the central nervous system [7, 8]. Studies have demonstrated the role of caffeine in improving cognitive skills, such as improving attention, increasing alertness rates, reducing tiredness and sleep duration [9, 10]. The amount of caffeine normally found in a cup of coffee can produce psychostimulant effects and increase the performance of individuals in clinical behavioral tasks [11]. However, there are hundreds of compounds in coffee, several of which with potential to contribute to coffee bioactivities directly or indirectly, for instance, some of them by interaction with caffeine [8, 9, 12].

The international coffee trade is concerned with only two coffee species: *Coffea arabica* L. and *Coffea canephora*. Both species proved to be sources of biologically active compounds, such as nicotinic acid, trigonelline, quinolinic acid, tannic acid, pyrogallic acid, chlorogenic acids and especially caffeine [10]. Considering this, and the fact that several coffee metabolites can act on the central nervous system or potentiate the effect of caffeine, it is extremely important to develop tools to gain a complete picture of the metabolites present in the whole biological matrix, and understand which compounds in the beverage are directly related to focus and diffuse attention when performing psychological tests.

Psychological attention tests are a way of examining the level of human attention. They are used in several situations, such as psychological diagnosis assessments, personnel selection, driving permission, and to assess information processing speed. Psychometric tests to check focus and diffuse attention are the most used in these cases. Focus attention is defined as the ability to select a source of information from all available at a given time and to be able to direct the attention (focus) to stimulus or tasks to be performed over time [13]. Diffuse attention is the mental function that focuses at the same time on various spatially dispersed stimuli, performing a quick collection of information and providing instant knowledge to the individual. Diffuse attention aims to investigate, evaluate and observe how quickly or slowly a person can discriminate against dispersed stimuli, such as a driver who drives on highways [14].

To understand which metabolites of coffee beverage can be related to human attention, a systematic method involving a wide variety of metabolites (metabolomic profiling) could be a very useful contribution. Nuclear Magnetic Resonance (NMR) spectroscopy is the analytical technique that can provide the most complete "metabolome" profile in a single analysis. ^{1}H NMR spectroscopy is rapid, reproducible, and stable over time, requiring a very simple sample preparation, and provides both qualitative and quantitative information about the metabolites present in a sample [15, 16]. Several studies have been carried out showing the application of ^{1}H NMR for the analysis of green and/or roasted coffee beans. For instance, this technique has been used to monitor changes in the composition of coffee during roasting [17, 18], to check adulteration in roasted coffee using corn, coffee husks, barley, and soybean [15], to discriminate coffee beans from different geographical origins [19], to differentiate coffee from different production systems [20], to evaluate the quality of green coffee or coffee beverages [17, 21], and to evaluate the anti-amyloidogenic properties of coffee and its constituents [22].

^{1}H NMR spectroscopy can provide the "metabolome", that is a chemical profile or fingerprint of whole tissues. Today, metabolomics constitutes a potent approach for the investigation and discovery of biomarkers in a large diversity of research domains [16, 22]. In this sense, NMR spectroscopy is a powerful tool capable of detecting a range of different types of metabolites simultaneously, providing valuable structural information with high reproducibility, although with low sensitivity (sub-millimolar concentrations) [23]. The fact that NMR has been routinely used for classical metabolic studies to characterize complex metabolite mixtures, has, in fact, made NMR the preferred technology in the field of metabolomics [24]. Thousands of metabolites in a single analysis are simultaneously monitored. The extraction of meaningful information from these large and complex datasets requires strategies, such as chemometrics tools that become essential for knowledge discovery in metabolomics [22]. In view of the above, in this study, we report the application of ^{1}H NMR coupled to PLS regression and VIP scores to identify the coffee metabolites potentially related to the human focus and diffuse attention. The success of a metabolomics approach requires variability of metabolites in the samples analyzed, to be able to reveal signals related to some activity. To obtain these variabilities, we used four different coffee blends that were offered to the volunteers before and after the psychological tests. In general, when applying the PLS model, greater variability in samples represents more interferents modeled, and then more robustness to the model, the inclusion of different coffee types, *i.e.*, a high chemical variation is fundamental to significantly correlate the psychological tests and the chemical data.

EXPERIMENTAL

Psychological Tests and Coffee Intake

Seventeen volunteers aged 18 to 52 years were selected to participate in this study (15 women and 02 men). Although the number of volunteers is not large, it is suitable for a pilot study, as previously reported [23 - 25]. Detailed information about the study was provided to the volunteers before the screening, and all volunteers gave their consent before starting the experiment. The study was carried out in accordance with the Ethics Committee in the Research from UTFPR – Paraná – Brazil, under protocol number 2.810.398.

The psychological tests were applied by a clinical psychologist. Before coffee consumption, the focus attention test (AC-vector) and diffuse attention test (TADIM) were applied. The AC-vector test consists of a sheet with 3 model symbols, 21 lines, each one with 21 symbols. In each horizontal line, 7 symbols (one-third of the total lines) must always be canceled according to the 3 model symbols at the top of the sheet [26]. In the TADIM test, 50 numbers are shown scattered on the sheet randomly, within a geometric shape. Over the course of 4 minutes, the task of the volunteer was to write down as many numbers as possible in the ordinal sequence. The volunteers must circulate the last number noted at the end of the 4 minutes [26].

After AC-vector and TADIM tests, the volunteers consumed a total of 40 mL of coffee consisting of 10 mL of each coffee blend: hard/rioysh medium roast (coffee blend 1), hard/rioysh dark roast (coffee blend 2), rioysh/rio medium-dark roast (coffee blend 3), and hard/Robusta light roast (coffee blend 4).

Commercial coffees were provided by a cooperative of coffee producers in the state of Paraná – Brazil. The samples were from the 2016/2017 and 2017/2018 crops. More details on the post-harvest processes of coffee cherries by this cooperative can be found in a study by Baqueta *et al*. [27]. Green coffee beans were roasted in an OPUS 40 industrial roaster (CIA Lilla, Brazil). Hard/rioysh medium roast blend was roasted at 240 °C for 15 min (coffee type 1), hard/rioysh dark roast blend (coffee type 2) at 260 °C for 15 min, rioysh/rio medium-dark roast blend (coffee type 3) at 255 °C for 15 min and hard/Robusta light roast blend (coffee type 4) at 210 °C for 20 min. The proportion of each final coffee blend was not provided by the cooperative due to commercial secrecy. The roasted beans were quickly cooled by spraying cold water (40 L) over the beans and poured into a cooling silo. Coffee beans were then ground with a coffee roller grinding machine (ORION, CIA Lilla – Brazil) with fine adjustment. All samples have the purity and quality seals from the Brazilian Association of Coffee Industries (ABIC). Coffee blend 1 has the traditional coffee seal, coffee blends 2 and 3 have the extra strong coffee seal and coffee blend 4 has the premium coffee seal.

Coffee brews were prepared through percolation, following the Brazilian recommendations, where 100 g of roasted and ground coffee was extracted with 1000 mL of hot mineral water at 92 °C [28]. The temperature was controlled using a digital thermometer (Incoterm - model 9791.16.1.00). The beverages were filtered through 103 paper filters (Mellita, Brazil) using a filter holder and a domestic pot. All coffees were stored in thermal bottles before serving (about 15 min after preparation).

The volunteers received coffee beverages of four different cup qualities in white plastic cups with random codes. Before drinking, they smelled each coffee beverage two to four times and chose the most preferred one based on the smell sensory attribute. Mineral water was served to the volunteers before and after coffee intake. The volunteers tested the samples from left to right and filled out an evaluation form (Fig. 1) indicating the order of preference of the coffee with respect to smell.

Psychological tests were again applied to measure focus and diffuse attention, before and after coffee intake. To avoid learning effects, different psychometric tests were used before and after drinking coffee. After drinking coffee, the focus attention test named TACOM-B and the diffuse attention test named TEDIF were applied. In the TACOM-B test, four models of traffic caution signs were presented to the volunteers. Below, on the same paper sheet, there are other traffic caution signs as well.

The task of the volunteers was to scratch every time they encountered the same model presented, and in this sense, the focus of their attention becomes the traffic caution signs, more especially those that were presented as models. The time for the TACOM-B test was 1 minute and 30 seconds [26, 29]. For the TEDIF test, 50 numbers were shown scattered on the sheet at random, within a geometric shape. Over the course of four minutes, the volunteer's task was to reach as many numbers as possible in the ordinal sequence. Every 1 min, under the command of the psychologist, the volunteer should circle the last number noted [26].

```
Name_____ Age____ Date__/__/____

You are getting coffee samples of four different cup qualities.
Please, taste the samples from left to right and make the order
of your preference in relation to the smell:

_____   _____   _____   _____
less preferred                          more preferred
```

Fig. (1). Evaluation form.

Sample Extraction for ¹H NMR Analysis

A ¹H NMR spectrum was obtained for the most preferred coffee of each volunteer. For these analyses, all reagents were purchased from Sigma Aldrich except deuterated solvents: H_2O-d_2 and 3-(trimethylsilyl)-propionic-2,2,3,3-d_4 acid sodium salt – TMSP purchased from Eurisotop. One hundred milligrams of roasted and ground coffee were extracted with 1.5 mL of phosphate buffer (90 mM, pH 6.0) in deuterated water containing 0.01% 3-(trimethylsilyl)-propionc-2,2,3,3-d_4 acid sodium salt (TMSP), as standard. The extraction took 1 h in a preheated bath maintained at a fixed temperature of 90 °C. Subsequently, aliquots of supernatant were collected with an automatic volumetric pipette and transferred to NMR tubes for analysis.

¹H NMR Analysis

¹H NMR spectra were recorded on a Varian 14.4 T NMR spectrometer, operating at a proton NMR frequency. The spectrometer was equipped with a High-Field Triple Resonance Probe, using deuterated water for the internal lock. For each sample, 256 scans were recorded with the following parameters: temperature of 298 K, relaxation delay of 2 s, acquisition time of 16 min, observed pulse of 5.80 µs, and spectral window of 16.00 ppm. A presaturation sequence was used to suppress the residual water signal at 4.83 ppm (power = 22 Hz, presaturation delay = 2 s).

The free induction decays (FIDs) of ^{1}H NMR spectra were Fourier transformed, and the resulting spectra were phased, baseline-corrected and calibrated to TMSP at 0.00 ppm. The spectral intensities were reduced to integrated regions of equal width (0.04 ppm) corresponding to the region of δ 0.00−12.00 with scaling on the standard at 0.00 ppm using the NMR Mestrenova software (Mestrelab Research, Spain). The region of δ 5.00−4.50, corresponding to the residual water signal, were excluded from the analysis. Binning was performed using Mestrenova software by normalizing for standard and each bin was the average sum of 0.04 ppm intervals.

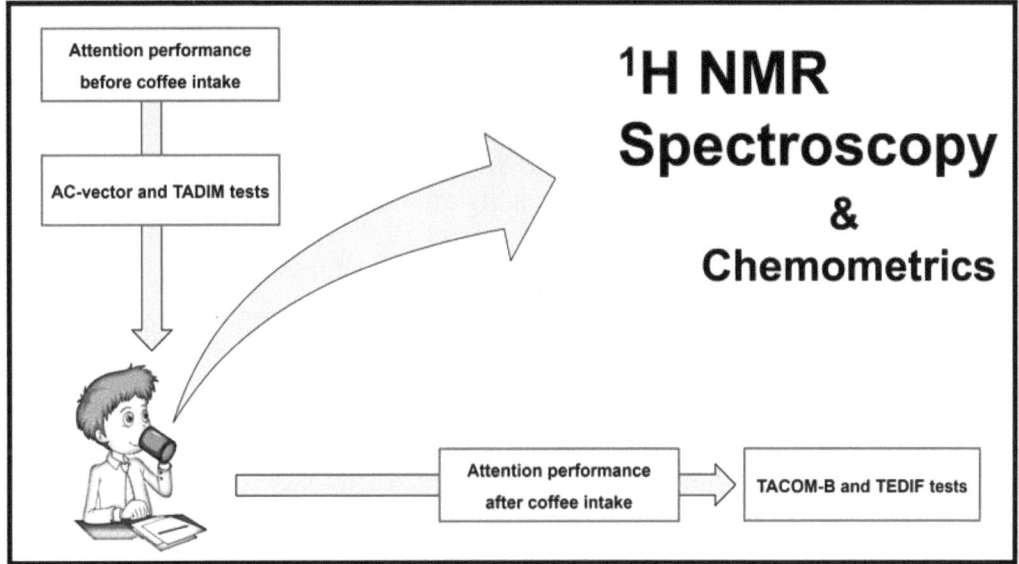

Fig. (2). Graphical representation of the experiment flow.

Chemometrics

To extract information from psychological tests related to coffee intake and spectral data from ^{1}H NMR, a chemometric data treatment was performed. A data matrix was built by integrating the ^{1}H NMR signals. The method used for modeling was the Partial Least Square (PLS) regression. We have used PLS regression in our previous studies [30, 31] and more information about this method can be found in the literature [32, 33]. In PLS regression, a dependent variable (**y**) is modeled using latent variables (LV), maximizing the covariance between an **X** matrix and a **y** vector. In our PLS models, ^{1}H NMR spectra represented the data matrix (**X**), and the psychological test results represented the vector (**y**). To determine which of the measured variables (spectral regions) contribute most to the definition of the PLS regression model, the Variable Importance in Projection (VIP) scores were used [34]. The VIP scores were calculated to determine which coffee compounds could be related to human focus and diffuse attention. A schematic representation of the steps of this study is shown in Fig. (2).

RESULTS AND DISCUSSIONS

Human Attention and Coffee Preference

The ages and individual scores in the psychological tests of each volunteer are presented in Table 1. Analyzing these results, it was possible to understand the changes in focus and diffuse attention in the volunteer's tests before and after coffee drinking. The performance results on psychological tests applied before and after coffee intake, related to the age of the volunteers, are presented below. Regarding the focus attention tests (AC-vector and TACOM-B), 88.2% of the volunteers had an increase in the test results after drinking coffee, while 11.8% showed a decrease or no change. Considering the age variable, the results showed that among volunteers aged 18 to 25 years old, 91.7% of them showed an increase in focus attention after drinking coffee. On the other hand, all volunteers aged 25 to 32 years old presented an increase in focus attention after coffee intake. Among the volunteers over 33 years old, 66.7% had an increase in focus attention, while 33.3% had a decrease in the performance of psychological tests after drinking coffee. In general, these results indicate that volunteers under the age of 33 performed better on psychological tests after coffee intake.

Table 1. Results of psychological tests applied before and after coffee intake.

		Before Coffee Intake		After Coffee Intake		Favorite Coffee
		AC-vector	TADIM	TACOM-B	TEDIF	
Volunteer	Age	Scores	Scores	Scores	Scores	
1	24	64	28	96	24	4
2	18	116	50	130	34	4
3	52	64	42	67	27	4
4	21	71	25	128	35	1
5	18	84	50	130	43	1
6	20	81	35	128	29	4
7	19	115	45	130	28	3
8	39	73	40	130	48	4
9	19	82	34	126	47	3

(Table 1) cont.....

		Before Coffee Intake		After Coffee Intake		Favorite Coffee
		AC-vector	TADIM	TACOM-B	TEDIF	
Volunteer	Age	Scores	Scores	Scores	Scores	
10	38	91	50	130	41	4
11	21	47	38	81	30	4
12	19	97	50	129	41	2
13	19	68	50	128	25	4
14	18	66	28	98	12	3
15	19	77	47	130	42	3
16	18	99	31	130	29	1
17	26	78	28	122	36	4

In the evaluation of diffuse attention (TADIM and TEDIF tests), the results showed that 52.9% of the volunteers had lower test results after coffee intake, while 23.5% had higher test results and 23.5% maintained similar results before and after drinking coffee. These results show that in more than half of the volunteers, coffee intake contributed to reducing diffuse attention, that is, to keep the focus on stimuli without dispersion. In fact, if diffuse attention decreases, the focus is maintained, corroborating to results obtained in the tests of focus attention, which showed an increase in focus attention in most volunteers. From a general check of the results of psychological tests, it is suggested that coffee intake attenuated diffuse attention and improved focus attention in most volunteers.

The coffee blends were classified by the smell from the most preferred to the least preferred. Considering the results obtained, it can be verified that coffee blend 4 was the most preferred one (52.9%), among volunteers, followed by coffee blend 3 (23.5%) and coffee blend 1 (17.6%). Coffee blend 2 was the least preferred (5.8%), probably associated to its strong flavor appearing to be slightly burnt due to characteristics of extra dark roast. Coffee blend 4 is classified as a premium coffee composed of selected beans with pleasant aroma and flavor that increases its sensory quality, which justifies the greatest preference of volunteers for this coffee blend. However, independent of the coffee preference all volunteers presented an improvement in focus attention, and it is hard to draw conclusions over the psychological test results only by analyzing, or trying to correlate, these results without a more accurate and effective chemical analytical evaluation.

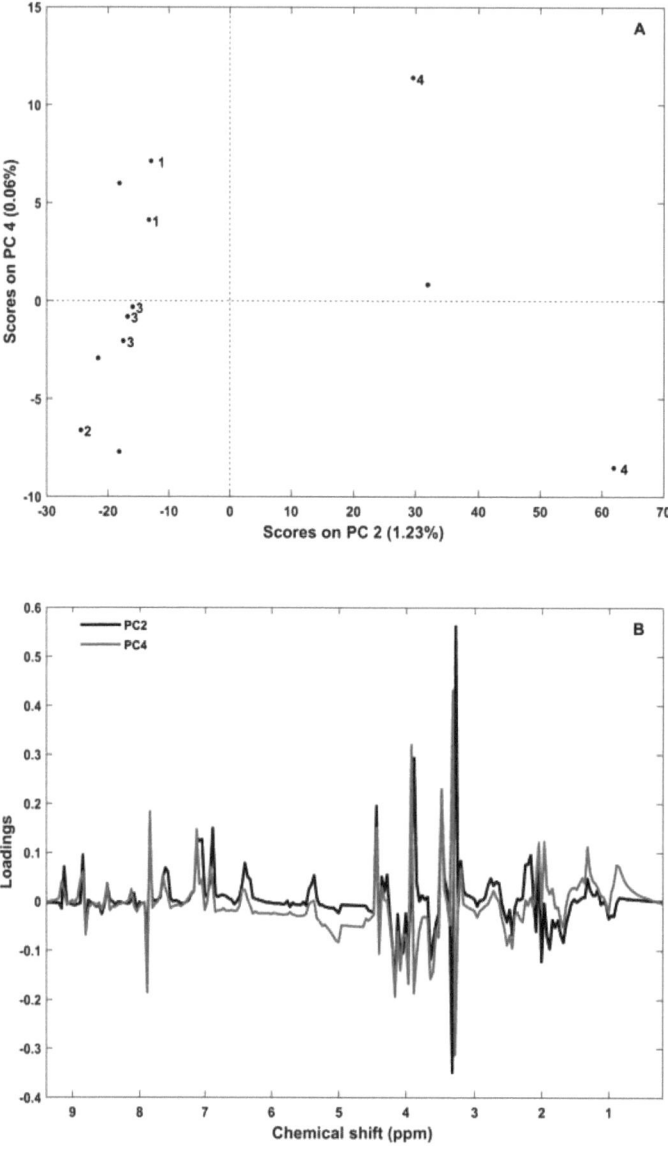

Fig. (3). PCA results. (**a**) Scores. (**b**) Loadings.

In order to verify that tested coffees were different, a Principal Component Analysis (PCA) was performed using three representative ^{1}H NMR spectra from each coffee blend. The PCA scores results (Fig. **3a**) show that there are differences between the coffee blends. However, a similarity was observed between coffee blends 2 and 3. In fact, these two coffee blends present a roasting degree more intense (dark and medium-dark) compared to the other blends (medium and light). Then, it is possible to conclude that the degrees of roasting produce a difference in coffee metabolites related to the smell that influences the preferences of the volunteers. The loadings plot (Fig. **3b**) shows a difference in intensity (*i.e.*, concentration of compounds) of the ^{1}H NMR signals, highlighting the intensities of the signals related to the smell at δ 3.28, 3.92, and 3.45 attributed to polysaccharides, and at δ 3.48 attributed to quinic acids. Considering that the polysaccharides after roasting add body and sweet notes to the beverage [1], and the degradation of quinic acids results in products that contribute to the aromatic fraction of coffee, with a predominance of phenolic derivatives [1], these results justify the differences achieved between coffee blends and different preferences of volunteers.

Even today, the chemistry of roasted coffee flavor is not fully understood. Previous studies have shown that polysaccharides, especially sucrose, the most abundant, act as aroma precursors, originating several compounds, including furans, aldehydes, carboxylic acids, *etc.*, which affect both the taste and aroma of the beverage. It is also known that chlorogenic acids (CGA) in coffee, among them caffeoylquinic acids (CQA), feruloylquinic acids (FQA) and dicaffeoylquinic acids (diCQA), are responsible for the formation of aroma, color and astringency of coffee, besides its contribution for bitterness when they are thermally degraded resulting in phenolic substances [35, 36].

Correlation between the Coffee Chemical Profile and Psychological Tests

Before carrying out the chemometric analysis, the ^{1}H NMR spectra were analyzed visually, in order to recognize the presence of dissimilarities arising from technical aspects, associated with factors such as shimming, signal suppression due to deuterated water, phase shifts, and baseline corrections. If these variations are present, they lead to errors and incorrect classification of samples in the chemometric analysis [15]. However, the ^{1}H NMR data did not show any visible problems related to these variables.

For the most appreciated coffee of each volunteer, a ^{1}H NMR spectrum was recorded. Fig. (**4**) shows the ^{1}H NMR profile of different coffee blends. Samples for ^{1}H NMR analysis were prepared reproducing the same conditions used to prepare the coffee brews in order to obtain a chemical profile as similar to that of coffee beverages as possible. The chemical composition of coffee extracts is very complex, showing that the main differences between the compositions of coffee blends were basically quantitative and not qualitative. In this sense, chemometric tools are critical to extract the important information correlating the ^{1}H NMR resonances to psychological tests.

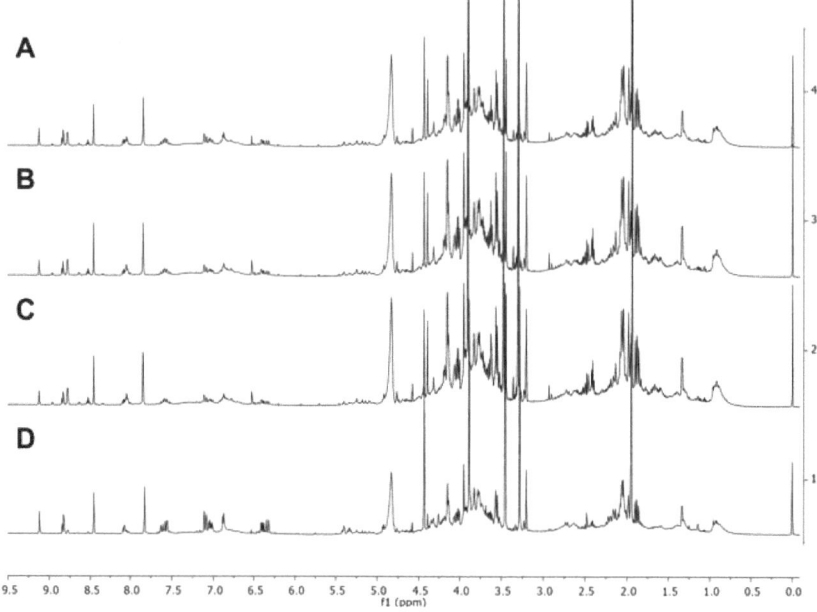

Fig. (4). ¹H NMR profiles of coffee beverages. Hard/rioysh medium-dark roast−Coffee blend 1 (**A**); Hard/rioysh dark roast−coffee blend 2 (**B**); rioysh/rio medium-dark roast−coffee blend 3 (**C**); and hard/Robusta light roast−coffee blend 4 (**D**).

To correlate the chemical profile of different coffee blends with the results of psychological tests, PLS regression (mean center preprocess, 11 latent variables - LVs for diffuse attention and 9 LVs for focus attention) was used, followed by interpretation of VIP scores in Fig. (**5**). Two PLS models were built to correlate the psychological tests to ¹H NMR profiles, one for focus attention test results (AC-vector and TACOM-B) and one for diffuse attention test results (TADIM and TEDIF). The important coffee compounds highlighted in the model were identified based on the ¹H NMR chemical shifts, coupling constants and comparison with the literature [17, 20, 22]. Some chemical resonances, in the region of δ 1.00 to 9.50, were important for both psychological attention models (focus and diffuse attention). The compounds identified for the highlighted signals were: trigonelline (δ 9.13), chlorogenic acids (δ 7.13), caffeine (δ 7.85, 3.89, 3.49, 3.29), acetate (δ 1.96), lipids (δ 0.93), lactate (δ 1.33), quinic acids (δ 4.17, 3.48, 2.08, 2.01, and 1.89), γ-quinide (δ 2.53) and polysaccharides (δ 5.20, 4.45, 4.16, 3.92, 3.80, 3.72, 3.42, and 3.28). Other metabolites were exclusively important to focus attention model: formate (δ 8.49), choline (δ 3.21), myo-inositol (δ 3.33), citrate and malate (δ 2.69).

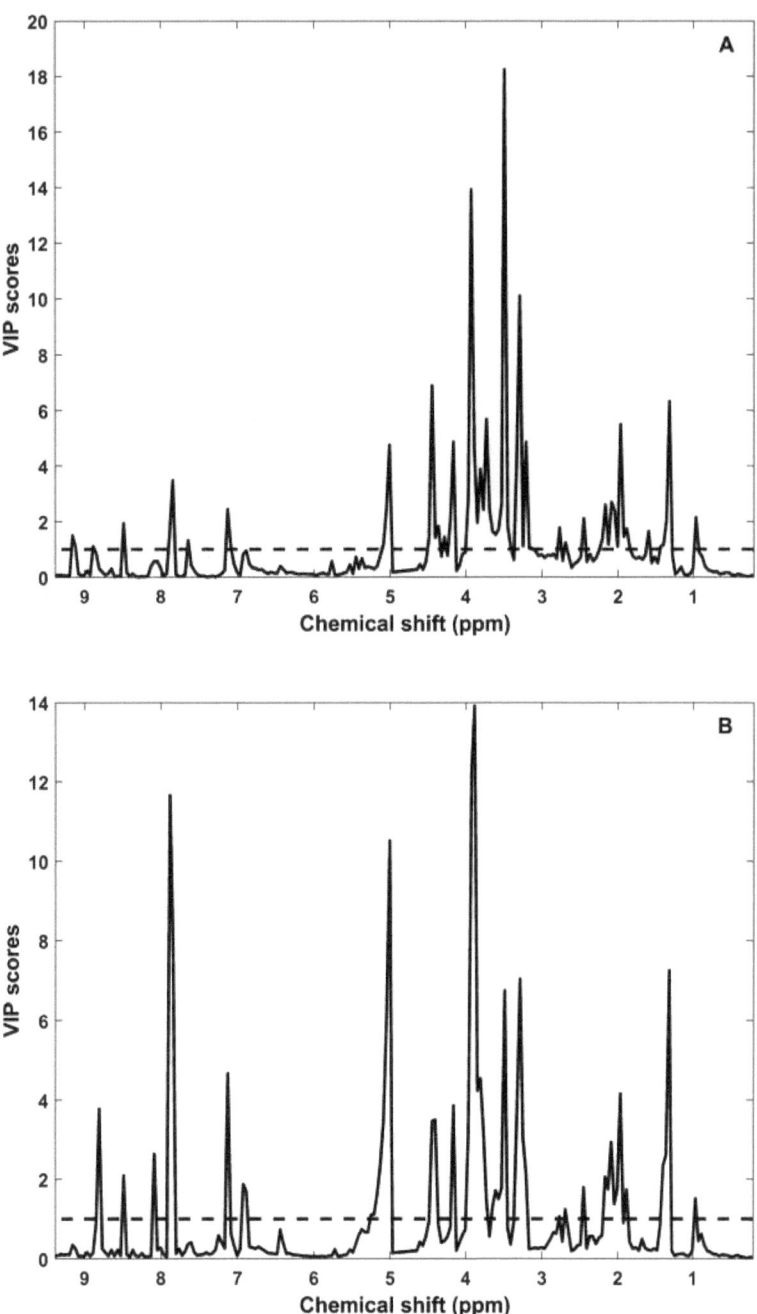

Fig. (5). VIP scores. (**A**) For the model built to focus attention. (**B**) For the model built to diffuse attention.

Although the main psychoactive effect of coffee has been attributed to caffeine [7, 8], our results showed that other compounds in coffee, such as formate, choline, myo-inositol, citrate, and malate are more related to the focus attention in volunteers after coffee intake. On the other hand, caffeine was more related to diffuse attention, corroborating the role of caffeine in improving attention, raising alertness levels and reducing tiredness, the latter also related to diffuse attention. In fact, studies have shown that the interaction of caffeine with other coffee compounds can produce beneficial and stimulating effects of coffee beverages, in addition to the isolated effect of caffeine [8, 9, 12].

From a scientific point of view, we recognize some limitations of this study that must be remembered in future research on the same topic. For example, this study used small groups of volunteers of different ages and, therefore, the conclusions may not be readily generalized for all age groups. An additional aspect that deserves attention for future research is the analysis of volatile compounds by gas chromatography to verify which compounds influence consumer preferences based on the coffee smell. Our findings may encourage new studies to further investigate deeply the relationship between coffee intake with focus and diffuse attention tests.

CONCLUDING REMARKS

In general, coffee intake attenuated diffuse attention and improved focus attention in most volunteers. In addition, the ^{1}H NMR profile proved to be an adequate analytical technique for an investigation of relevant coffee compounds related to human attention. The metabolites that contributed for both diffuse and focus attention were caffeine, trigonelline, chlorogenic acids, acetate, lipids, lactate, γ-quinide, and polysaccharides. The metabolites exclusively important to focus attention were formate, choline, myo-inositol, citrate, and malate. Therefore, the ^{1}H NMR profile associated with chemometric tools is interesting to assess the correlation between coffee compounds and human attention.

CONSENT FOR PUBLICATION

Not applicable.

CONFLICT OF INTEREST

The authors confirm that this chapter contents have no conflict of interest

ACKNOWLEDGEMENTS

The authors thank "Coordenação de Aperfeiçoamento de Pessoal de Nível Superior (CAPES)" for the master scholarship of Michel R. Baqueta. Patrícia

Valderrama thanks "Fundação Araucária" (process 033/2019).

REFERENCES

[1] Toledo, P.R.A.B.; Pezza, L.; Pezza, H.R.; Toci, A.T. Relationship between the different aspects related to coffee quality and their volatile compounds. *Compr. Rev. Food Sci. Food Saf.*, **2016**, *15*, 705-719.
[http://dx.doi.org/10.1111/1541-4337.12205]

[2] Mancini, R.S.; Wang, Y.; Weaver, D.F. Phenylindanes in brewed coffee inhibit amyloid-beta and tau aggregation. *Front. Neurosci.*, **2018**, *12*, 735.
[http://dx.doi.org/10.3389/fnins.2018.00735] [PMID: 30369868]

[3] Rodríguez-Artalejo, F.; López-García, E. Coffee consumption and cardiovascular disease: A Condensed review of epidemiological evidence and mechanisms. *J. Agric. Food Chem.*, **2018**, *66*(21), 5257-5263.
[http://dx.doi.org/10.1021/acs.jafc.7b04506] [PMID: 29276945]

[4] Jiang, X.; Zhang, D.; Jiang, W. Coffee and caffeine intake and incidence of type 2 diabetes mellitus: a meta-analysis of prospective studies. *Eur. J. Nutr.*, **2014**, *53*(1), 25-38.
[http://dx.doi.org/10.1007/s00394-013-0603-x] [PMID: 24150256]

[5] Andersen, L.F.; Jacobs, D.R., Jr; Carlsen, M.H.; Blomhoff, R. Consumption of coffee is associated with reduced risk of death attributed to inflammatory and cardiovascular diseases in the Iowa Women's Health Study. *Am. J. Clin. Nutr.*, **2006**, *83*(5), 1039-1046.
[http://dx.doi.org/10.1093/ajcn/83.5.1039] [PMID: 16685044]

[6] de Mendonça, A.; Cunha, R.A. Therapeutic opportunities for caffeine in Alzheimer's disease and other neurodegenerative disorders. *J. Alzheimers Dis.*, **2010**, *20* Suppl. 1, S1-S2.
[http://dx.doi.org/10.3233/JAD-2010-01420] [PMID: 20448305]

[7] Angeloni, G.; Guerrini, L.; Masella, P.; Bellumori, M.; Daluiso, S.; Parenti, A.; Innocenti, M. What kind of coffee do you drink? An investigation on effects of eight different extraction methods. *Food Res. Int.*, **2019**, *116*, 1327-1335.
[http://dx.doi.org/10.1016/j.foodres.2018.10.022] [PMID: 30716922]

[8] Bae, J-H.; Park, J-H.; Im, S.S.; Song, D.K. Coffee and health. *Integr. Med. Res.*, **2014**, *3*(4), 189-191.
[http://dx.doi.org/10.1016/j.imr.2014.08.002] [PMID: 28664096]

[9] Haskell-Ramsay, C.F.; Jackson, P.A.; Forster, J.S.; Dodd, F.L.; Bowerbank, S.L.; Kennedy, D.O. The acute effects of caffeinated black coffee on cognition and mood in healthy young and older adults. *Nutrients*, **2018**, *10*(10)E1386
[http://dx.doi.org/10.3390/nu10101386] [PMID: 30274327]

[10] Borota, D.; Murray, E.; Keceli, G.; Chang, A.; Watabe, J.M.; Ly, M.; Toscano, J.P.; Yassa, M.A. Post-study caffeine administration enhances memory consolidation in humans. *Nat. Neurosci.*, **2014**, *17*(2), 201-203.
[http://dx.doi.org/10.1038/nn.3623] [PMID: 24413697]

[11] Childs, E.; de Wit, H. Subjective, behavioral, and physiological effects of acute caffeine in light, nondependent caffeine users. *Psychopharmacology (Berl.)*, **2006**, *185*(4), 514-523.
[http://dx.doi.org/10.1007/s00213-006-0341-3] [PMID: 16541243]

[12] Hečimović, I.; Belščak-Cvitanović, A.; Horžić, D.; Komes, D. Comparative study of polyphenols and caffeine in different coffee varieties affected by the degree of roasting. *Food Chem.*, **2011**, *129*(3), 991-1000.
[http://dx.doi.org/10.1016/j.foodchem.2011.05.059] [PMID: 25212328]

[13] Benczik, E.B.P.; Leal, G.C.; Cardoso, T. A utilização do teste de atenção concentrada (AC) para a população infanto-juvenil: uma contribuição para a avaliação neuropsicológica. *Rev Psicopedag*, **2016**, *33*, 37-49.

[14] Marín Rueda, F.J. Desempenho no teste de atenção dividida como resultado da idade das pessoas. *Estud. Psicol.,* **2011**, *28*, 251-259.
[http://dx.doi.org/10.1590/s0103-166x2011000200012]

[15] de Moura Ribeiro, M.V.; Boralle, N.; Redigolo Pezza, H.; Pezza, L.; Toci, A.T. Authenticity of roasted coffee using ^{1}H NMR spectroscopy. *J. Food Compos. Anal.,* **2017**, *57*, 24-30.
[http://dx.doi.org/10.1016/j.jfca.2016.12.004]

[16] Choi, Y.H.; Sertic, S.; Kim, H.K.; Wilson, E.G.; Michopoulos, F.; Lefeber, A.W.M.; Erkelens, C.; Prat Kricun, S.D.; Verpoorte, R. Classification of Ilex species based on metabolomic fingerprinting using nuclear magnetic resonance and multivariate data analysis. *J. Agric. Food Chem.,* **2005**, *53*(4), 1237-1245.
[http://dx.doi.org/10.1021/jf0486141] [PMID: 15713047]

[17] Wei, F.; Furihata, K.; Miyakawa, T.; Tanokura, M. A pilot study of NMR-based sensory prediction of roasted coffee bean extracts. *Food Chem.,* **2014**, *152*, 363-369.
[http://dx.doi.org/10.1016/j.foodchem.2013.11.161] [PMID: 24444949]

[18] Ciampa, A.; Renzi, G.; Taglienti, A.; Sequi, P.; Valentini, M. Studies on coffee roasting process by means of nuclear magnetic resonance spectroscopy. *J. Food Qual.,* **2010**, *33*, 199-211.
[http://dx.doi.org/10.1111/j.1745-4557.2010.00306.x]

[19] Consonni, R.; Cagliani, L.R.; Cogliati, C. NMR based geographical characterization of roasted coffee. *Talanta,* **2012**, *88*, 420-426.
[http://dx.doi.org/10.1016/j.talanta.2011.11.010] [PMID: 22265520]

[20] Consonni, R.; Polla, D.; Cagliani, L.R. Organic and conventional coffee differentiation by NMR spectroscopy. *Food Control,* **2018**, *94*, 284-288.
[http://dx.doi.org/10.1016/j.foodcont.2018.07.013]

[21] Kwon, D-J.; Jeong, H-J.; Moon, H.; Kim, H-N.; Cho, J-H.; Lee, J-E. Assessment of green coffee bean metabolites dependent on coffee quality using a ^{1}H NMR-based metabolomics approach. *Food Res. Int.,* **2015**, *67*, 175-182.
[http://dx.doi.org/10.1016/j.foodres.2014.11.010]

[22] Ciaramelli, C.; Palmioli, A.; De Luigi, A.; Colombo, L.; Sala, G.; Riva, C.; Zoia, C.P.; Salmona, M.; Airoldi, C. NMR-driven identification of anti-amyloidogenic compounds in green and roasted coffee extracts. *Food Chem.,* **2018**, *252*, 171-180.
[http://dx.doi.org/10.1016/j.foodchem.2018.01.075] [PMID: 29478529]

[23] Gant, N.; Ali, A.; Foskett, A. The influence of caffeine and carbohydrate coingestion on simulated soccer performance. *Int. J. Sport Nutr. Exerc. Metab.,* **2010**, *20*(3), 191-197.
[http://dx.doi.org/10.1123/ijsnem.20.3.191] [PMID: 20601736]

[24] Songisepp, E.; Hütt, P.; Rätsep, M.; Shkut, E.; Kõljalg, S.; Truusalu, K.; Stsepetova, J.; Smidt, I.; Kolk, H.; Zagura, M.; Mikelsaar, M. Safety of a probiotic cheese containing Lactobacillus plantarum Tensia according to a variety of health indices in different age groups. *J. Dairy Sci.,* **2012**, *95*(10), 5495-5509.
[http://dx.doi.org/10.3168/jds.2011-4756] [PMID: 22863096]

[25] Beltrán-Barrientos, L.M.; González-Córdova, A.F.; Hernández-Mendoza, A.; Torres-Inguanzo, E.H.; Astiazarán-García, H.; Esparza-Romero, J.; Vallejo-Cordoba, B. Randomized double-blind controlled clinical trial of the blood pressure-lowering effect of fermented milk with Lactococcus lactis: A pilot study. *J. Dairy Sci.,* **2018**, *101*(4), 2819-2825.
[http://dx.doi.org/10.3168/jds.2017-13189] [PMID: 29428751]

[26] Hood P. Segredos dos Psicotécnicos para quem não quer ser surpreendido. *Testes de Atenção*: Teste d2 – AC-Vetor – TECON – AD e AS – TADIM e TEDIF – TADIS e TACOM – TEALT, TEADI e TEACO-FF – BPA. **2018**.

[27] Baqueta, M.R.; Do Prado Silva, J.T.; Moya Moreira, T.F.; Canesin, E.A.; Gonçalves, O.H.; Dos

Santos, A.R. Extração e caracterização de compostos do resíduo vegetal casca de café. *Brazilian J Food Res,* **2017**, *8*, 68.
[http://dx.doi.org/10.3895/rebrapa.v8n2.6887]

[28] Brazil. *Ministério da Agricultura P e A. Instrução Normativa no 8, de 11 de junho de 2003,* **2003**.

[29] Hood P. Segredos dos Psicotécnicos para quem não quer ser surpreendido. *Testes de Atenção Discriminativa*: TADIS (1 e 2), Concentrada: TACOM (A e B), e Concentrada Complexa: TACOM (C e D). **2018**.

[30] Baqueta, M.R.; Coqueiro, A.; Valderrama, P. Brazilian Coffee Blends: A Simple and fast method by near-infrared spectroscopy for the determination of the sensory attributes elicited in professional coffee cupping. *J. Food Sci.,* **2019**, *84*(6), 1247-1255.
[http://dx.doi.org/10.1111/1750-3841.14617] [PMID: 31116425]

[31] Baqueta, M.R.; Coqueiro, A.; Março, P.H.; Valderrama, P. Quality Control parameters in the roasted coffee industry: a proposal by using microNIR spectroscopy and multivariate calibration. *Food Anal. Methods,* **2020**, *13*, 50-60.
[http://dx.doi.org/10.1007/s12161-019-01503-w]

[32] Ribeiro, J.S.; Ferreira, M.M.C.; Salva, T.J.G. Chemometric models for the quantitative descriptive sensory analysis of Arabica coffee beverages using near infrared spectroscopy. *Talanta,* **2011**, *83*(5), 1352-1358.
[http://dx.doi.org/10.1016/j.talanta.2010.11.001] [PMID: 21238720]

[33] Ribeiro, J.S.; Augusto, F.; Salva, T.J.G.; Thomaziello, R.A.; Ferreira, M.M.C. Prediction of sensory properties of Brazilian Arabica roasted coffees by headspace solid phase microextraction-gas chromatography and partial least squares. *Anal. Chim. Acta,* **2009**, *634*(2), 172-179.
[http://dx.doi.org/10.1016/j.aca.2008.12.028] [PMID: 19185116]

[34] Vitale, R.; Bevilacqua, M.; Bucci, R.; Magrì, A.D.; Magrì, A.L.; Marini, F. A rapid and non-invasive method for authenticating the origin of pistachio samples by NIR spectroscopy and chemometrics. *Chemom. Intell. Lab. Syst.,* **2013**, *121*, 90-99.
[http://dx.doi.org/10.1016/j.chemolab.2012.11.019]

[35] Farah, A.; Monteiro, M.C.; Calado, V.; Franca, A.S.; Trugo, L.C. Correlation between cup quality and chemical attributes of Brazilian coffee. *Food Chem.,* **2006**, *98*, 373-380.
[http://dx.doi.org/10.1016/j.foodchem.2005.07.032]

[36] Franca, A.S.; Mendonça, J.C.F.; Oliveira, S.D. Composition of green and roasted coffees of different cup qualities. *Lebensm. Wiss. Technol.,* **2005**, *38*, 709-715.
[http://dx.doi.org/10.1016/j.lwt.2004.08.014]

CHAPTER 3

NMR Spectroscopy for Probing the Structural Determinants of Aptamer Optimization and Riboswitch Engineering

B. Bora[1], Ö. Uğurlu[1], E. Man[1], M. Gültan[1], C. Özyurt[2] and S. Evran[1,*]

[1] *Department of Biochemistry, Faculty of Science, Ege University, 35100 Bornova-İzmir, Turkey*

[2] *Department of Chemistry and Chemical Processing Technologies, Lapseki Vocational School, Canakkale Onsekiz Mart University, Canakkale, Lapseki, Turkey*

Abstract: Nucleic acid aptamers are single-stranded DNA or RNA molecules that can fold into unique conformations and specifically recognize various targets, such as small molecules, proteins, cells, and tissues. Aptamers are selected *in vitro* through an iterative process called Systematic Evolution of Ligands by Exponential Enrichment (SELEX). As aptamers possess several advantages over antibodies, several diagnostic and therapeutic applications have emerged in recent years. Aptamers also attract interest as they form the receptor domain of RNA-based riboswitches that function as natural modulators of gene expression. Aptamer domain of riboswitch can sense the metabolite and this binding event is transduced into a conformational change, thereby transcriptional or translational control is achieved. Riboswitch engineering has gained importance due to the potential use of artificial riboswitches in biosensors and next-generation therapeutics. Therefore, understanding the structural basis of ligand binding and conformational change is critical for the success of optimization or re-engineering of aptamers. Since crystallization of aptamer-small molecule target complexes is particularly difficult, NMR provides an indispensable tool for structural analysis. In this chapter, we first give a brief information about aptamers and riboswitches. Then, we review the NMR structures of aptamers and riboswitches reported to date. We highlight the importance of NMR for identification of ligand binding mechanism, post-SELEX optimization of aptamers, as well as for the design of artificial riboswitches. In this context, we also give some examples of aptamer studies involving a combination of NMR and other techniques.

Keywords: Aptamer, Aptamer-ligand interaction, NMR-guided design, Riboswitch.

---

[*] **Corresponding author Serap Evran:** Department of Biochemistry, Faculty of Science, Ege University, 35100 Bornova-İzmir, Turkey; Tel: + 90 232 3112304; Fax: +90 232 3115485; E-mail: serap.evran@gmail.com

Atta-ur-Rahman and M. Iqbal Choudhary (Eds.)
All rights reserved-© 2020 Bentham Science Publishers

INTRODUCTION

Selection of Aptamers

Aptamers are single-stranded DNA or RNA oligonucleotides that can bind to their targets with high affinity and specificity due to their ability to form specific three-dimensional structures [1]. Aptamers are selected by using an *in vitro* process called SELEX (Systematic Evolution of Ligands by EXponential Enrichment) [2, 3]. The SELEX process consists of three basic steps: binding, elution and amplification (Fig. 1). In the first step, the target of interest is incubated with initial DNA or RNA library. Initial library of a typical SELEX process consists of about 10^{13} to 10^{15} different sequences. The library is composed of chemically synthesized oligonucleotides that contain a random region of 20 to 100 nucleotides, which are flanked by specific primer binding sites at the 5' and 3' ends. Following incubation of the target and the library at pre-defined conditions, unbound sequences are removed by washing with buffer. The binding sequences are then amplified by polymerase chain reaction (PCR) *via* the primer binding sites. With repetitive cycles of binding, elution and amplification, the initial random pool of oligonucleotides is reduced to the enriched sequences that show the highest affinity and specificity to the target molecule. Binding properties of the selected aptamers depend on the molecular structure of the target, design of the initial random library, selection conditions, and the ratio of the target to the library. Typically, gradually increasing stringent conditions are applied to obtain aptamers with high affinity and specificity. The stringency can be achieved by reducing target concentration, increasing the number of wash steps, or by decreasing incubation time. Enrichment of high-affinity sequences indicates that the SELEX process can be finalized. The enriched pool is then sequenced and the aptamer candidates are chemically synthesized for further characterization of binding properties.

Aptamers bind to their targets through non-covalent interactions such as van der Waals forces, hydrogen bonding and electrostatic interactions [4]. Binding affinity and specificity of aptamers are comparable to antibodies. Moreover, aptamers are superior to antibodies due to their small size, stability, low cost, ease of chemical synthesis and unlimited target range [5]. Aptamers targeting small molecules, metal ions, peptides, proteins and cells can be developed *in vitro* by excluding the need for living systems. With these properties, aptamers have enormous potential to be used in therapeutics and diagnostics [6, 7]. Aptamers can be designed for many different purposes, such as modulating the immune system, inhibiting enzyme activity, drug transport, and blocking receptor binding [8].

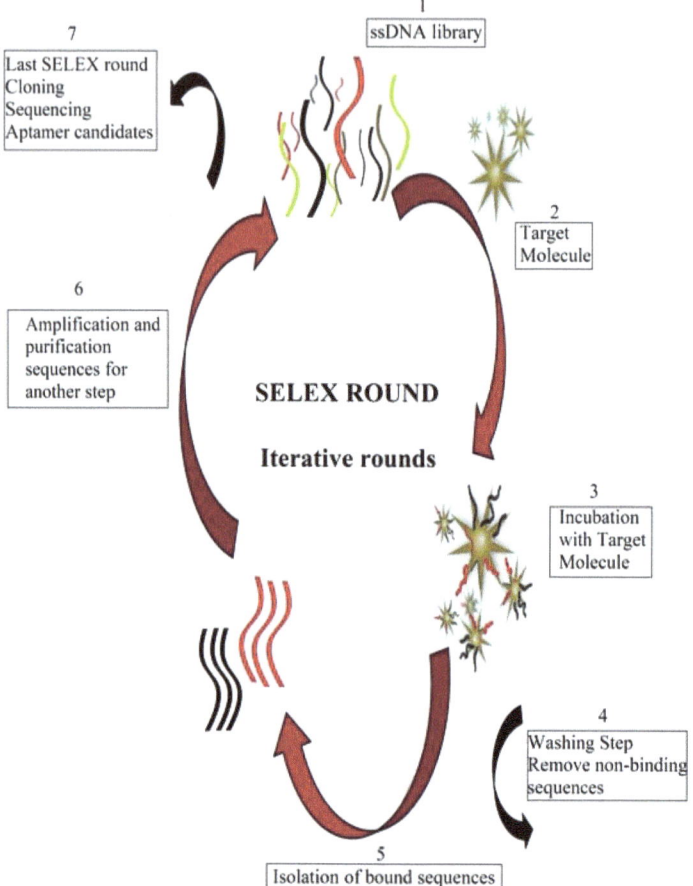

Fig. (1). Schematic representation of SELEX.

Post-SELEX Modifications of Aptamers

Therapeutic application of aptamers is usually limited by short half-lives due to rapid degradation by nucleases. For instance, the half-life of a 16-mer oligonucleotide in rat plasma is less than 1 minute [9]. The oligonucleotide that is promising as an anticoagulant has a limited half-life of 108 seconds [10]. Hence, post-SELEX optimization is a powerful approach to overcome therapeutic limitations since it allows modifications to the aptamer structure to improve stability and binding properties [11 - 16].

Truncation is one of the post-SELEX optimization strategies that relies on shortening the aptamer by removing the nucleotides that are not involved in target binding. The constant primer binding sites of aptamers were shown to contribute minimally to the overall structure [17]. Hence, primer binding sequences can be considered for truncation. The shortened aptamers may have a better or same affinity against the target molecule [18]. The truncated RNA aptamer has been shown to bind to human interleukin 8 with a substantial improvement in K_d value [19]. The previously developed APT[STX1] aptamer [20] has been subjected to truncation and site-directed mutagenesis, yielding a 30-fold higher affinity against saxitoxin [18].

One advantage of aptamers is that they can be easily chemically modified for post-SELEX purposes. Modifications can be introduced into the sugar ring, nucleobase, and the phosphodiester linkage. Modification on RNA aptamers is usually performed at 2' position of nucleosides, while modifications on DNA aptamers are performed at the phosphodiester backbone. Chemical modifications make the aptamers resistant against nuclease degradation, enhancing the therapeutic usage of aptamers [21]. The mirror image of an aptamer, also known as Spiegelmer has been shown to overcome the nuclease degradation problem [22 - 29].

2'-Amino modification is another strategy to improve the binding properties of aptamers. 2'-amino modified aptamers are selected from the RNA library with the replacement of 2'-hyroxyl group with 2'-amino group of pyrimidines. The 2'-amino modified aptamer has been found to bind to thyroid-stimulating hormone with a K_d value of 2.5 nM and high specificity. Moreover, 300-fold higher serum stability has been achieved [30]. 2'-Fluoro modification is another approach to enhance the thermal stability of aptamers [31]. In addition, it lowers the cost of synthesis since 2'-fluoro pyrimidine does not require the protection/deprotection step [32]. The 2'-fluoro and 2'-amino modified aptamers, as well as the aptamers containing both modifications, have been compared in terms of K_d values and half-lives (Table 1) [33]. 2'-Fluoro modification has been shown to provide thermal stability [34, 35]. 2'-Methoxy modification has made the aptamer more resistant against renal and vaginal nucleases [36]. Hence, a combination of 2'-fluoro and 2'-methoxy modifications has been revealed to be a more effective way to enhance nuclease resistance and stability [37, 38]. In addition to those modifications, 4'-thio [39] and 2'-fluoroarabinonucleotide [40] modifications have been shown to enhance the half-life and K_d value, respectively.

Table 1. Comparison of the 2'-fluoro and 2'-amino modified aptamers.

Feature	2'-Fluoro Modification	2'-NH$_2$ Modification	Both modifications
K_d	6.8 nM	1.8 nM	106 nM
Half-life	6 hours	80 hours	48 hours

SOMAmers (Slow off-rate modified aptamers) show improved binding affinity and kinetics compared to traditional aptamers. Besides, they show resistance to nucleases [12]. Most SOMAmer modifications are made through uridine at the 5-position [12]. The library containing uridine modified with 5-benzylaminocarbonyl, 5- tryptaminocarbonyl, 5- naphthylmethylaminocarbonyl, 5- tyrosylaminocarbonyl, 5-phenylethyl-1-aminocarbonyl, or 5-(2- naphthylmethly) aminocarbonyl has been used to develop SOMAmer against toxin A and B [41].

In addition to the modifications introduced to sugar ring or bases, several aptamers include modifications at the phosphodiester bond [42, 43]. Phosphodiester bond can be replaced by methylphosphonate or phosphothionate linkage [44 - 46], which may provide higher resistance in serum.

Capping of aptamers from the ends is also a useful strategy to provide nuclease resistance and to increase the half-life. 3'-end capping can be achieved with biotin, biotin-streptavidin, or inverted thymine. The aptamers in clinical trials are mostly optimized by 3'-end capping strategy [47, 48]. 3'- Biotin also provides resistance against 3'-exonucleases [49]. Moreover, 3'-end biotin-streptavidin capping has been shown to decelerate the rate of renal clearance *in vivo* [50]. Aptamers can also be capped with cholesterol, polyethylene glycol (PEG), proteins, amine or phosphate at their 5'-ends [51]. An increase in half-life has been achieved by the conjugation of aptamer with cholesterol [50]. Upon PEG modification, the half-life of MP7 aptamer has been dramatically increased from 1 hour to 24-48 hours [52, 53].

Riboswitches

Genetic regulation through RNA is a common strategy used by bacteria. Riboswitches are RNA sequences ranging from 35 to 200 nucleotides, which act as specific ligand-specific receptors. Riboswitches recognize several metabolites such as coenzymes, small molecules, magnesium ion, nucleotide derivatives, enzyme cofactors, signaling molecules and amino acid residues. Upon binding to their ligands, riboswitches can regulate gene expression through conformational changes [54].

Riboswitches were first discovered in 2002 as RNA-based intracellular sensors of vitamin derivatives in bacteria [55]. The first riboswitch has been revealed to regulate the production of enzymes involved in vitamin B1 (thiamine) biosynthesis in *Escherichia coli*. Subsequently, several riboswitches were characterized in the archaea and eukaryotes like fungi and plants. Riboswitches are classified into families and classes based on their ligand specificity and secondary structure [56]. Riboswitches consist of two different regions named aptamer domain and expression platform [57]. When the ligand concentration reaches a certain level, the aptamer domain binds the ligand. Then, the expression domain (platform) located close to the aptamer domain undergoes conformational changes. Depending on the conformational change, gene expression can be either initiated or repressed (Fig. **2**). Transcription, translation, splicing and mRNA stability are regulated by riboswitch mechanisms. Although riboswitches usually have a negative effect on gene expression, in some cases, riboswitches exert their effect by accelerating gene expression [58].

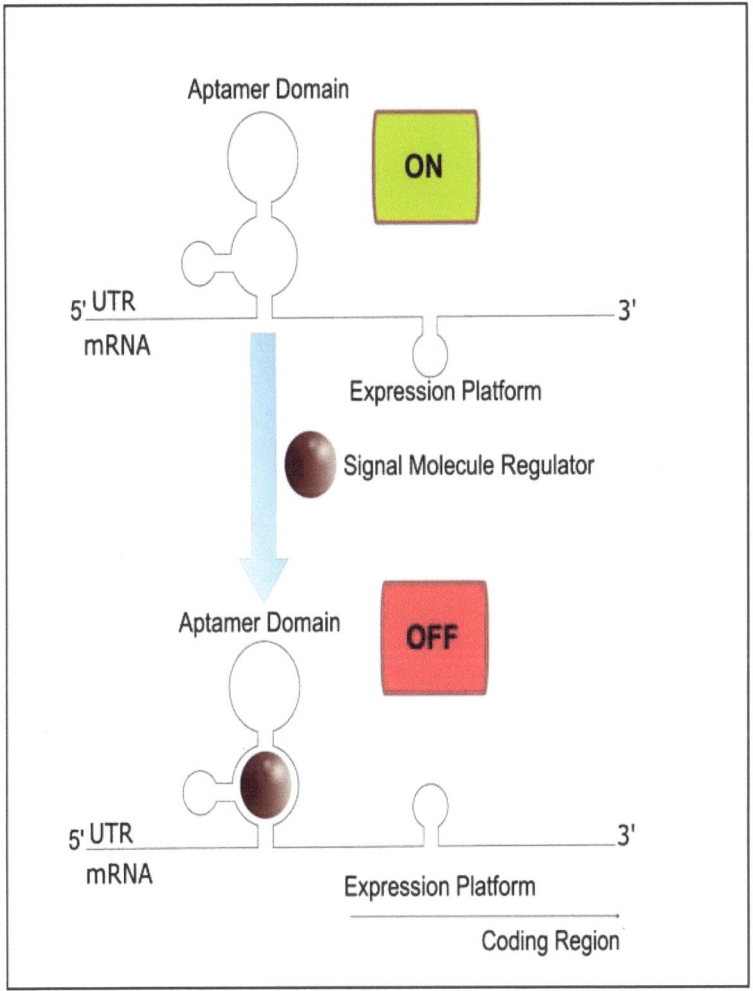

Fig. (2). Schematic representation of the riboswitch regulation mechanism.

Riboswitches do not only bind a specific ligand, but also trigger gene expression.

Therefore, riboswitches display strong potential for the design of novel biosensors [59]. Riboswitches are also novel targets for antibiotic development [60, 61]. Riboswitch mechanism has been successfully used to control gene expression in human pathogen *Mycobacterium tuberculosis* [62].

NMR AND COMPLEMENTARY TECHNIQUES FOR APTAMERS AND RIBOSWITCHES

The structures of aptamers and riboswitches attract much interest to reveal interactions with their targets, as well as to perform post-SELEX optimization or design of artificial riboswitches. For this aim, three-dimensional structures of aptamers can be solved by using either X-ray crystallography or NMR spectroscopy. Since the crystallization of nucleic acids is more difficult [63, 64], NMR is a faster method to gain insight into aptamer and aptamer-small target molecule complex. NMR has been performed to reveal the structural properties of aptamers in the presence and absence of targets. The changes in the imino spectrum of nucleic acids provide more information about their bound and non-bound conformational structures [64, 65]. However, NMR also has disadvantages. Due to the overlapping and expansion of NMR signals, the determination of high-resolution structures of large nucleic acids (above 200 nucleotides) by NMR is limited [66, 67]. Therefore, crystal structures of many aptamer-large target molecule complexes have been solved by X-ray crystallography [68, 69].

A combination of small angle X-ray scattering (SAXS) and NMR is significant for determination of the global structure of larger RNA and RNA complexes [63, 65]. Haller *et al.* (2011) studied the ligand-induced folding process of S-adenosylmethionine type II (SAM-II) riboswitch using NMR and single-molecule fluorescence imaging [70]. According to their findings, the riboswitch adopts a more flexible structure in the absence of Mg^{2+} ions and its ligand. Unfolded riboswitch conformations and transitions between the hairpin and pseudoknot-like states have been observed by smFRET analysis. In the presence of Mg^{2+}, pseudoknot-like structure that resembles the global fold of ligand-bound structure has been observed by both NMR and smFRET. Structural forms of SAM-II riboswitch under different folding conditions have been characterized by NMR spectroscopy combined with SAXS. In that study, NMR has provided more information to understand the nature of secondary and tertiary structural regulation of RNA in the presence of Mg^{2+} ion and the metabolite. Based on the SAXS data, a computational analysis has been performed regarding the four different folding states of the aptamer in order to better understand the conformational changes [71]. Computational approaches as well as biophysical and biochemical tools could give more knowledge about the dynamics of riboswitches and aptamers. In addition, they are useful to estimate potential binding sites and the efficient binding forces between aptamers and targets [72 - 74].

Single-molecule Förster resonance energy transfer (smFRET) is a significant tool that can be used in combination with NMR. smFRET is useful for real-time monitoring of the three-dimensional structural dynamics and interactions of riboswitches and aptamers from sub-seconds to minutes [70, 75, 76]. However, NMR analysis can reveal interaction dynamics in atomic resolution on a time scale from picoseconds to seconds [65, 67, 77]. The three-dimensional structure of adenine riboswitch (Asw) models (apoA, apoB, holo) from *Vibrio vulnificus* has been investigated by NMR and smFRET to investigate the effects of adenine and Mg^{2+} conditions on the three-state conformational balance in both secondary and tertiary structure. smFRET analysis has revealed substates of apoA form, which could not be detected by NMR analysis. That finding has been explained by the limitations of NMR of RNA such that imino resonances can only be observed for structured RNA elements, and that smFRET analysis requires labeling and tethering of RNA. The authors conclude that NMR and smFRET data highlight the necessity to integrate complementary solution structural techniques to uncover full conformational space of riboswitch RNA [78].

Several studies show that a combination of NMR and other methods gives more detailed information about the binding mechanisms of aptamers [79 - 81]. As shown in Table **2**, riboswitch structures have been resolved through NMR or its combination with different techniques. Likewise, NMR has been used to reveal several aptamer structures. Table **3** summarizes the aptamer structures resolved by NMR in the last 15 years.

NMR as well as isothermal titration calorimetry (ITC), semi-elastic light scattering (QELS) and small-angle X-ray spectroscopy (SAXS) methods have been used to study the binding mechanism of DNA aptamer [80]. Thermodynamic parameters, such as dissociation constant (K_D), Gibbs free energy (ΔG), enthalpy (ΔH) and entropy change (ΔS) and stochiometry can be obtained by isothermal titration calorimetry (ITC). Thus, a combination of NMR and ITC can provide information about the high-affinity binding mechanisms of aptamers. Likewise, surface plasmon resonance (SPR) can give additional data on thermodynamic parameters [4, 82].

Table 2. The riboswitch structures resolved by NMR to date.

Riboswitch	Technique	PDB ID	Reference
SAM-II	NMR and Fluorescence Spectroscopy	2QWY	[70]
Neomycin sensing riboswitch	NMR	2MXS	[83]
add Adenine	X-Ray, NMR	5SWE 5E54 1Y26	[84]
Synthetic neomycin riboswitch	Solution NMR	2N0J, 2KXM 2MXS	[74, 83]
Fluoride riboswitch	Solution NMR	5KH8	[85]
L-Glutamine	X-Ray, NMR, MD	5DDP	[73]
Mg^{2+} riboswitch	Solution NMR	2LI4	
preQ1 Class I	Solution NMR	2L1V	[88]
preQ1 Class II	Solution NMR	2MIY	[89]
SAM/SAH-binding riboswitch	Solution NMR	6HAG	[90]

Neomycin-sensing riboswitch is known for its high substrate specificity. Even a single functional group can be discriminated by riboswitch as in the case of paromomycin and neomycin. Duchard-Ferner *et al.* (2016) have studied neomycin-sensing riboswitch by solution NMR in order to explore underlying conformational dynamics that causes high substrate specificity [83]. Ligand-bound solution structures have been observed to be almost identical. However, the regulatory function is completely different. Consistently, the authors have reported a 500 times higher affinity for neomycin compared to paromomycin. It has been shown that the absence of a single amino group in paromomycin results in a more flexible riboswitch-ligand complex. Lack of intermolecular interaction as a result of lacking amino group has been attributed to the destabilized OFF state. Even though the same structural behaviour has been observed by solution NMR data, the authors have concluded that discriminative intermolecular interactions are a determinant of regulatory function. As a supporting study, molecular dynamics (MD) simulations have been performed to explore details of the structural dynamics of the aminoglycoside-N1-riboswitch complexes [74]. Consistently, a more flexible complex with paromomycin has been confirmed by MD simulation data. The more flexible riboswitch-ligand complex for paromomycin has been attributed to the lack of contact between distinctive 6'-OH group of paromomycin and G9 phosphate of the riboswitch. Their study contributes to understanding the importance and complementary role of MD simulations in the effort of investigating structural and dynamic behaviour of riboswitches.

Table 3. Aptamer structures resolved by NMR in the last 15 years.

Aptamer Name	Structure	Target	PDB Code
TMR	RNA	TAMRA	6GZK 6GZR
Thrombin binding aptamer (TBA)	2'-F-ANA/DNA Chimeric	Thrombin	5MJX
11-mer-3'-t-TBA	Truncated DNA	Thrombin	2MKO 2MKM
Fucose-TBA	Fucose conjugated DNA	Thrombin	2LYG
TBA	5'-5' inversion modified DNA	Thrombin	2IDN
GTP binding class II RNA aptamer	RNA	GTP	5LWJ
Class I GTP Aptamer	RNA	GTP	2AU4
Ribosomal protein S8 binding aptamer	RNA	*B. anthracis* ribosomal protein S8	2LUN
VEGF aptamer	LNA Modified DNA	VEGF	2M53
MUC1 aptamer	DNA Truncated 23mer	Mucin 1	2L5K
RNA aptamer against AML1 Runt domain	RNA	AML1 Runt Domain	2RRC
RNA aptamer against prion protein	RNA	Prion protein	2RSK 6K84
Aptamer R12	RNA	A partial peptide of a prion protein	2RU7
TMR Aptamer	RNA	Tetramethylrhodamine	6GZR 6GZK
OBA	DNA	Ochratoxin	6J2W
RNA aptamer against bovine prion protein	RNA	Bovine prion protein	2RQJ
HIV TAR RNA binding RNA aptamer	RNA	HIV TAR RNA	2RN1
HIV-1 TAR RNA binding aptamer	LNA/RNA	HIV-1 TAR RNA	2OOM 2PN9
Anti-NFkB RNA aptamer	RNA	NFkB	2JWV

Chemical exchange saturation transfer (CEST) NMR spectroscopy has been applied in order to understand the structural dynamics of *Bacillus cereus* fluoride riboswitch [85]. Fluoride riboswitch is responsible for the regulation of toxicity response. The holo-like apo structure of the riboswitch is critical to understand conformational dynamics of the riboswitch. Considering that there is only a marginal gap between its binding affinity (K_d ~0.1 mM) and the toxic threshold of fluoride (MIC ~1 mM), exploring the 'hidden' difference in conformational dynamics between the apo and holo aptamers is very important. It has been shown that *Bacillus cereus* fluoride riboswitch aptamer adopts identical tertiary structures in solution with and without ligand. Using CEST-NMR technique, the authors have revealed that the structured ligand-free aptamer transiently accesses a low-populated (~1%) and short-lived (~3 ms) excited conformational state that unravels a conserved 'linchpin' base pair for signal transcription termination. Upon fluoride binding, this highly localized fleeting process is allosterically suppressed to activate transcription.

It has been shown that U14C substitution on neomycin sensing riboswitch does not alter the structure. In order to test this negligible structural effect, Krepl *et al.* (2018) used MD simulations and revealed a structure-stabilizing monovalent cation binding site in wild-type RNA [86]. They also used NMR analysis in order to prove the stabilizing effect of this monovalent cation binding site. As a result, the NMR experiments have confirmed a strong response to changes in potassium concentration associated with the U14 nucleotide of wild type RNA. Contrarily, almost no such change has been observed for the C14+ nucleotide. Even in the presence of Mg^{2+}, potassium ion has been observed to be bound in the vicinity of U14. Thus, the authors have suggested that this monovalent cation-binding site may be characteristic of the U-turn motif structure in general. As supported by quantum mechanic analysis and NMR data, MD simulations have clearly revealed a stronger interaction for the C14+ system. The authors have concluded that the presence of the U14 cation-binding site in the wild-type structure reduces the differences in the local interaction networks between the U14 and C14+ systems. Thus, the interactions after the U14C+ substitution mimic the native U14 system much more closely than expected. That study contributes to the importance of a combination of MD and NMR analysis to reveal unidentified features of riboswitches, which could be underreported in the X-ray structures.

Determination of ligand binding affinities *in vitro* may not always be a promising factor for riboswitch activity *in vivo*. The differences in the ionic and molecular composition may affect the *in vitro* and *in vivo* behaviour of riboswitch. For this reason, Kim *et al.* (2017) have studied the 2'-deoxyguanosine binding riboswitch in metabolic context [87]. For this purpose, they have prepared *Bacillus subtilis* cell lysate in order to mimic *in vivo* conditions. The importance of guanosine binding to 2'-deoxyguanosine riboswitch has been revealed by in-cell NMR. However, it has been shown that a mutated version of an aptamer has almost no ability to discriminate cognate ligand and guanosine. Six analogs have been screened and it has been shown that guanosine could interfere with the ligand-binding pocket, thus lowering the affinity between the aptamer and cognate ligand. The authors have concluded that this finding is in line with the existing crystal structure of the wild-type aptamer domain in complex with guanosine, as well as the 2'-deoxyguanosine-bound X-ray structure. Since the mutated version aptamer lacks specificity to guanine-based nucleotides, the authors conclude that a pool of metabolites could still bind to the mutated aptamer in cells due to high concentrations of each metabolite. That study presents a particular example where *in vitro* data differ from the data obtained under in-cell experiment setups. That study clearly indicates the importance of in-cell like environment for a thorough characterization of riboswitch behaviour.

CONCLUSION

Aptamers and riboswitches are powerful tools for biosensor design and therapeutic applications. Understanding the structural and regulatory principles of natural riboswitches and binding mechanism of aptamers is critical for post-SELEX optimization, as well as to design synthetic riboswitches that can respond to a particular ligand. Structural and biophysical approaches are required to gain insight into the three-dimensional structures and binding mechanisms of aptamers. Moreover, these methods are of great importance for investigating the control mechanisms of gene expression of riboswitches. Unfortunately, a single method is not sufficient to clarify the structural dynamics and transitions. Therefore, there is a need for combined approaches that can provide complementary information. In recent years, combinations of various approaches have been performed to investigate aptamer–target interactions and riboswitch mechanisms. As summarized in this chapter, there are several studies that show the power of NMR in combination with other experimental and computational approaches. The results obtained from the structural studies would give important clues for better understanding how aptamers and riboswitches bind to their ligands with high affinity and specificity.

CONSENT FOR PUBLICATION

Not applicable.

CONFLICT OF INTEREST

The authors confirm that this chapter contents have no conflict of interest.

ACKNOWLEDGEMENTS

We acknowledge the fellowships of Ezgi Man and Özge Uğurlu from the Republic of Turkey, Ministry of Development (Grant number 2016K121190).

REFERENCES

[1] Gelinas, A.D.; Davies, D.R.; Janjic, N. Embracing proteins: structural themes in aptamer-protein complexes. *Curr. Opin. Struct. Biol.,* **2016**, *36*, 122-132.
[http://dx.doi.org/10.1016/j.sbi.2016.01.009] [PMID: 26919170]

[2] Tuerk, C.; Gold, L. Systematic evolution of ligands by exponential enrichment: RNA ligands to bacteriophage T4 DNA polymerase. *Science,* **1990**, *249*(4968), 505-510.
[http://dx.doi.org/10.1126/science.2200121] [PMID: 2200121]

[3] Ellington, A.D.; Szostak, J.W. In vitro selection of RNA molecules that bind specific ligands. *Nature,* **1990**, *346*(6287), 818-822.
[http://dx.doi.org/10.1038/346818a0] [PMID: 1697402]

[4] Cai, S.; Yan, J.; Xiong, H.; Liu, Y.; Peng, D.; Liu, Z. Investigations on the interface of nucleic acid aptamers and binding targets. *Analyst (Lond.),* **2018**, *143*(22), 5317-5338.
[http://dx.doi.org/10.1039/C8AN01467A] [PMID: 30357118]

[5] Zhang, Y.; Lai, B.S.; Juhas, M. Recent advances in aptamer discovery and applications. *Molecules,* **2019**, *24*(5), 941.
[http://dx.doi.org/10.3390/molecules24050941] [PMID: 30866536]

[6] Kaur, H.; Bruno, J.G.; Kumar, A.; Sharma, T.K. Aptamers in the therapeutics and diagnostics pipelines. *Theranostics,* **2018**, *8*(15), 4016-4032.
[http://dx.doi.org/10.7150/thno.25958] [PMID: 30128033]

[7] Wang, T.; Chen, C.; Larcher, L.M.; Barrero, R.A.; Veedu, R.N. Three decades of nucleic acid aptamer technologies: Lessons learned, progress and opportunities on aptamer development. *Biotechnol. Adv.,* **2019**, *37*(1), 28-50.
[http://dx.doi.org/10.1016/j.biotechadv.2018.11.001] [PMID: 30408510]

[8] Ismail, S.I.; Alshaer, W. Therapeutic aptamers in discovery, preclinical and clinical stages. *Adv. Drug Deliv. Rev.,* **2018**, *134*, 51-64.
[http://dx.doi.org/10.1016/j.addr.2018.08.006] [PMID: 30125605]

[9] de Smidt, P.C.; Le Doan, T.; de Falco, S.; van Berkel, T.J. Association of antisense oligonucleotides with lipoproteins prolongs the plasma half-life and modifies the tissue distribution. *Nucleic Acids Res.,* **1991**, *19*(17), 4695-4700.
[http://dx.doi.org/10.1093/nar/19.17.4695] [PMID: 1891360]

[10] Griffin, L.C.; Tidmarsh, G.F.; Bock, L.C.; Toole, J.J.; Leung, L.L. *In vivo* anticoagulant properties of a novel nucleotide-based thrombin inhibitor and demonstration of regional anticoagulation in extracorporeal circuits. *Blood,* **1993**, *81*(12), 3271-3276.
[http://dx.doi.org/10.1182/blood.V81.12.3271.3271] [PMID: 8507864]

[11] Nallagatla, S.R.; Heuberger, B.; Haque, A.; Switzer, C. Combinatorial synthesis of thrombin-binding aptamers containing iso-guanine. *J. Comb. Chem.,* **2009**, *11*(3), 364-369.
[http://dx.doi.org/10.1021/cc800178m] [PMID: 19243167]

[12] Gold, L.; Ayers, D.; Bertino, J.; Bock, C.; Bock, A.; Brody, E.N.; Carter, J.; Dalby, A.B.; Eaton, B.E.; Fitzwater, T.; Flather, D.; Forbes, A.; Foreman, T.; Fowler, C.; Gawande, B.; Goss, M.; Gunn, M.; Gupta, S.; Halladay, D.; Heil, J.; Heilig, J.; Hicke, B.; Husar, G.; Janjic, N.; Jarvis, T.; Jennings, S.; Katilius, E.; Keeney, T.R.; Kim, N.; Koch, T.H.; Kraemer, S.; Kroiss, L.; Le, N.; Levine, D.; Lindsey, W.; Lollo, B.; Mayfield, W.; Mehan, M.; Mehler, R.; Nelson, S.K.; Nelson, M.; Nieuwlandt, D.; Nikrad, M.; Ochsner, U.; Ostroff, R.M.; Otis, M.; Parker, T.; Pietrasiewicz, S.; Resnicow, D.I.; Rohloff, J.; Sanders, G.; Sattin, S.; Schneider, D.; Singer, B.; Stanton, M.; Sterkel, A.; Stewart, A.; Stratford, S.; Vaught, J.D.; Vrkljan, M.; Walker, J.J.; Watrobka, M.; Waugh, S.; Weiss, A.; Wilcox, S.K.; Wolfson, A.; Wolk, S.K.; Zhang, C.; Zichi, D. Aptamer-based multiplexed proteomic technology for biomarker discovery. *PLoS One,* **2010**, *5*(12): e15004.
[http://dx.doi.org/10.1371/journal.pone.0015004] [PMID: 21165148]

[13] Mendelboum Raviv, S.; Horváth, A.; Aradi, J.; Bagoly, Z.; Fazakas, F.; Batta, Z.; Muszbek, L.; Hársfalvi, J. 4-thio-deoxyuridylate-modified thrombin aptamer and its inhibitory effect on fibrin clot formation, platelet aggregation and thrombus growth on subendothelial matrix. *J. Thromb. Haemost.,* **2008**, *6*(10), 1764-1771.
[http://dx.doi.org/10.1111/j.1538-7836.2008.03106.x] [PMID: 18665927]

[14] Phillips, J.A.; Liu, H.; O'Donoghue, M.B.; Xiong, X.; Wang, R.; You, M.; Sefah, K.; Tan, W. Using azobenzene incorporated DNA aptamers to probe molecular binding interactions. *Bioconjug. Chem.,* **2011**, *22*(2), 282-288.
[http://dx.doi.org/10.1021/bc100402p] [PMID: 21247152]

[15] Eaton, B.E.; Gold, L.; Hicke, B.J.; Janjić, N.; Jucker, F.M.; Sebesta, D.P.; Tarasow, T.M.; Willis, M.C.; Zichi, D.A. Post-SELEX combinatorial optimization of aptamers. *Bioorg. Med. Chem.,* **1997**, *5*(6), 1087-1096.
[http://dx.doi.org/10.1016/S0968-0896(97)00044-8] [PMID: 9222502]

[16] Radom, F.; Jurek, P.M.; Mazurek, M.P.; Otlewski, J.; Jeleń, F. Aptamers: molecules of great potential. *Biotechnol. Adv.,* **2013**, *31*(8), 1260-1274.
[http://dx.doi.org/10.1016/j.biotechadv.2013.04.007] [PMID: 23632375]

[17] Cowperthwaite, M.C.; Ellington, A.D. Bioinformatic analysis of the contribution of primer sequences to aptamer structures. *J. Mol. Evol.,* **2008**, *67*(1), 95-102.
[http://dx.doi.org/10.1007/s00239-008-9130-4] [PMID: 18594898]

[18] Zheng, X.; Hu, B.; Gao, S.X.; Liu, D.J.; Sun, M.J.; Jiao, B.H.; Wang, L.H. A saxitoxin-binding aptamer with higher affinity and inhibitory activity optimized by rational site-directed mutagenesis and truncation. *Toxicon,* **2015**, *101*, 41-47.
[http://dx.doi.org/10.1016/j.toxicon.2015.04.017] [PMID: 25937337]

[19] Sung, H.J.; Choi, S.; Lee, J.W.; Ok, C.Y.; Bae, Y.S.; Kim, Y.H.; Lee, W.; Heo, K.; Kim, I.H. Inhibition of human neutrophil activity by an RNA aptamer bound to interleukin-8. *Biomaterials,* **2014**, *35*(1), 578-589.
[http://dx.doi.org/10.1016/j.biomaterials.2013.09.107] [PMID: 24129312]

[20] Handy, S.M.; Yakes, B.J.; DeGrasse, J.A.; Campbell, K.; Elliott, C.T.; Kanyuck, K.M.; Degrasse, S.L. First report of the use of a saxitoxin-protein conjugate to develop a DNA aptamer to a small molecule toxin. *Toxicon,* **2013**, *61*, 30-37.
[http://dx.doi.org/10.1016/j.toxicon.2012.10.015] [PMID: 23142073]

[21] Abeydeera, N.D.; Egli, M.; Cox, N.; Mercier, K.; Conde, J.N.; Pallan, P.S.; Mizurini, D.M.; Sierant, M.; Hibti, F.E.; Hassell, T.; Wang, T.; Liu, F.W.; Liu, H.M.; Martinez, C.; Sood, A.K.; Lybrand, T.P.; Frydman, C.; Monteiro, R.Q.; Gomer, R.H.; Nawrot, B.; Yang, X. Evoking picomolar binding in RNA by a single phosphorodithioate linkage. *Nucleic Acids Res.,* **2016**, *44*(17), 8052-8064.
[http://dx.doi.org/10.1093/nar/gkw725] [PMID: 27566147]

[22] Vater, A.; Klussmann, S. Turning mirror-image oligonucleotides into drugs: the evolution of Spiegelmer(®) therapeutics. *Drug Discov. Today,* **2015**, *20*(1), 147-155.
[http://dx.doi.org/10.1016/j.drudis.2014.09.004] [PMID: 25236655]

[23] Hoellenriegel, J.; Zboralski, D.; Maasch, C.; Rosin, N.Y.; Wierda, W.G.; Keating, M.J.; Kruschinski, A.; Burger, J.A. The Spiegelmer NOX-A12, a novel CXCL12 inhibitor, interferes with chronic lymphocytic leukemia cell motility and causes chemosensitization. *Blood,* **2014**, *123*(7), 1032-1039.
[http://dx.doi.org/10.1182/blood-2013-03-493924] [PMID: 24277076]

[24] Klussmann, S.; Nolte, A.; Bald, R.; Erdmann, V.A.; Fürste, J.P. Mirror-image RNA that binds D-adenosine. *Nat. Biotechnol.,* **1996**, *14*(9), 1112-1115.
[http://dx.doi.org/10.1038/nbt0996-1112] [PMID: 9631061]

[25] Nolte, A.; Klussmann, S.; Bald, R.; Erdmann, V.A. F€urste, J.P. Mirrordesign of L-oligonucleotide ligands binding to L-arginine. *Nat. Biotechnol.,* **1996**, *14*, 116-119.
[http://dx.doi.org/10.1038/nbt0996-1116]

[26] Williams, K.P.; Liu, X.H.; Schumacher, T.N.M.; Lin, H.Y.; Ausiello, D.A.; Kim, P.S.; Bartel, D.P. Bioactive and nuclease-resistant L-DNA ligand of vasopressin. *Proc. Natl. Acad. Sci. USA*, **1997**, *94*(21), 11285-11290.
[http://dx.doi.org/10.1073/pnas.94.21.11285] [PMID: 9326601]

[27] Wlotzka, B.; Leva, S.; Eschgfäller, B.; Burmeister, J.; Kleinjung, F.; Kaduk, C.; Muhn, P.; Hess-Stumpp, H.; Klussmann, S. In vivo properties of an anti-GnRH Spiegelmer: an example of an oligonucleotide-based therapeutic substance class. *Proc. Natl. Acad. Sci. USA*, **2002**, *99*(13), 8898-8902.
[http://dx.doi.org/10.1073/pnas.132067399] [PMID: 12070349]

[28] Ni, S.; Yao, H.; Wang, L.; Lu, J.; Jiang, F.; Lu, A.; Zhang, G. Chemical Modifications of Nucleic Acid Aptamers for Therapeutic Purposes. *Int. J. Mol. Sci.*, **2017**, *18*(8), 1683.
[http://dx.doi.org/10.3390/ijms18081683] [PMID: 28767098]

[29] Lin, Y.; Qiu, Q.; Gill, S.C.; Jayasena, S.D. Modified RNA sequence pools for in vitro selection. *Nucleic Acids Res.*, **1994**, *22*(24), 5229-5234.
[http://dx.doi.org/10.1093/nar/22.24.5229] [PMID: 7529404]

[30] Lin, Y.; Nieuwlandt, D.; Magallanez, A.; Feistner, B.; Jayasena, S.D. High-affinity and specific recognition of human thyroid stimulating hormone (hTSH) by *in vitro*-selected 2'-amino-modified RNA. *Nucleic Acids Res.*, **1996**, *24*(17), 3407-3414.
[http://dx.doi.org/10.1093/nar/24.17.3407] [PMID: 8811096]

[31] Adler, A.; Forster, N.; Homann, M.; Göringer, H.U. Post-SELEX chemical optimization of a trypanosome-specific RNA aptamer. *Comb. Chem. High Throughput Screen.*, **2008**, *11*(1), 16-23.
[http://dx.doi.org/10.2174/138620708783398331] [PMID: 18220540]

[32] Ruckman, J.; Green, L.S.; Beeson, J.; Waugh, S.; Gillette, W.L.; Henninger, D.D.; Claesson-Welsh, L.; Janjić, N. 2'-Fluoropyrimidine RNA-based aptamers to the 165-amino acid form of vascular endothelial growth factor (VEGF165). Inhibition of receptor binding and VEGF-induced vascular permeability through interactions requiring the exon 7-encoded domain. *J. Biol. Chem.*, **1998**, *273*(32), 20556-20567.
[http://dx.doi.org/10.1074/jbc.273.32.20556] [PMID: 9685413]

[33] Kubik, M.F.; Bell, C.; Fitzwater, T.; Watson, S.R.; Tasset, D.M. Isolation and characterization of 2'-fluoro-, 2'-amino-, and 2'-fluoro-/amino-modified RNA ligands to human IFN-gamma that inhibit receptor binding. *J. Immunol.*, **1997**, *159*(1), 259-267.
[PMID: 9200462]

[34] Pagratis, N.C.; Bell, C.; Chang, Y.F.; Jennings, S.; Fitzwater, T.; Jellinek, D.; Dang, C. Potent 2'-amino-, and 2'-fluoro-2'-deoxyribonucleotide RNA inhibitors of keratinocyte growth factor. *Nat. Biotechnol.*, **1997**, *15*(1), 68-73.
[http://dx.doi.org/10.1038/nbt0197-68] [PMID: 9035109]

[35] Tucker, C.E.; Chen, L.S.; Judkins, M.B.; Farmer, J.A.; Gill, S.C.; Drolet, D.W. Detection and plasma pharmacokinetics of an anti-vascular endothelial growth factor oligonucleotide-aptamer (NX1838) in rhesus monkeys. *J. Chromatogr. B Biomed. Sci. Appl.*, **1999**, *732*(1), 203-212.
[http://dx.doi.org/10.1016/S0378-4347(99)00285-6] [PMID: 10517237]

[36] Moore, M.D.; Cookson, J.; Coventry, V.K.; Sproat, B.; Rabe, L.; Cranston, R.D.; McGowan, I.; James, W. Protection of HIV neutralizing aptamers against rectal and vaginal nucleases: implications for RNA-based therapeutics. *J. Biol. Chem.*, **2011**, *286*(4), 2526-2535.
[http://dx.doi.org/10.1074/jbc.M110.178426] [PMID: 21106536]

[37] Lee, J.H.; Canny, M.D.; De Erkenez, A.; Krilleke, D.; Ng, Y.S.; Shima, D.T.; Pardi, A.; Jucker, F. A therapeutic aptamer inhibits angiogenesis by specifically targeting the heparin binding domain of VEGF165. *Proc. Natl. Acad. Sci. USA*, **2005**, *102*(52), 18902-18907.
[http://dx.doi.org/10.1073/pnas.0509069102] [PMID: 16357200]

[38] Kong, H.Y.; Byun, J. Nucleic Acid aptamers: new methods for selection, stabilization, and application

in biomedical science. *Biomol. Ther. (Seoul),* **2013**, *21*(6), 423-434.
[http://dx.doi.org/10.4062/biomolther.2013.085] [PMID: 24404332]

[39] Kato, Y.; Minakawa, N.; Komatsu, Y.; Kamiya, H.; Ogawa, N.; Harashima, H.; Matsuda, A. New NTP analogs: the synthesis of 4'-thioUTP and 4'-thioCTP and their utility for SELEX. *Nucleic Acids Res.,* **2005**, *33*(9), 2942-2951.
[http://dx.doi.org/10.1093/nar/gki578] [PMID: 15914669]

[40] Alves Ferreira-Bravo, I.; Cozens, C.; Holliger, P.; DeStefano, J.J. Selection of 2'-deoxy--'-fluoroarabinonucleotide (FANA) aptamers that bind HIV-1 reverse transcriptase with picomolar affinity. *Nucleic Acids Res.,* **2015**, *43*(20), 9587-9599.
[http://dx.doi.org/10.1093/nar/gkv1057] [PMID: 26476448]

[41] Ochsner, U.A.; Katilius, E.; Janjic, N. Detection of Clostridium difficile toxins A, B and binary toxin with slow off-rate modified aptamers. *Diagn. Microbiol. Infect. Dis.,* **2013**, *76*(3), 278-285.
[http://dx.doi.org/10.1016/j.diagmicrobio.2013.03.029] [PMID: 23680240]

[42] Yamamoto, T.; Nakatani, M.; Narukawa, K.; Obika, S. Antisense drug discovery and development. *Future Med. Chem.,* **2011**, *3*(3), 339-365.
[http://dx.doi.org/10.4155/fmc.11.2] [PMID: 21446846]

[43] Wang, R.E.; Wu, H.; Niu, Y.; Cai, J. Improving the stability of aptamers by chemical modification. *Curr. Med. Chem.,* **2011**, *18*(27), 4126-4138.
[http://dx.doi.org/10.2174/092986711797189565] [PMID: 21838692]

[44] Saccà, B.; Lacroix, L.; Mergny, J.L. The effect of chemical modifications on the thermal stability of different G-quadruplex-forming oligonucleotides. *Nucleic Acids Res.,* **2005**, *33*(4), 1182-1192.
[http://dx.doi.org/10.1093/nar/gki257] [PMID: 15731338]

[45] Zaitseva, M.; Kaluzhny, D.; Shchyolkina, A.; Borisova, O.; Smirnov, I.; Pozmogova, G. Conformation and thermostability of oligonucleotide d(GGTTGGTGTGGTTGG) containing thiophosphoryl internucleotide bonds at different positions. *Biophys. Chem.,* **2010**, *146*(1), 1-6.
[http://dx.doi.org/10.1016/j.bpc.2009.09.011] [PMID: 19846249]

[46] Pozmogova, G.E.; Zaitseva, M.A.; Smirnov, I.P.; Shvachko, A.G.; Murina, M.A.; Sergeenko, V.I. Anticoagulant effects of thioanalogs of thrombin-binding DNA-aptamer and their stability in the plasma. *Bull. Exp. Biol. Med.,* **2010**, *150*(2), 180-184.
[http://dx.doi.org/10.1007/s10517-010-1099-5] [PMID: 21240367]

[47] Fine, S.L.; Martin, D.F.; Kirkpatrick, P. Pegaptanib sodium. *Nat. Rev. Drug Discov.,* **2005**, *4*(3), 187-188.
[http://dx.doi.org/10.1038/nrd1677] [PMID: 15770779]

[48] Gupta, S.; Hirota, M.; Waugh, S.M.; Murakami, I.; Suzuki, T.; Muraguchi, M.; Shibamori, M.; Ishikawa, Y.; Jarvis, T.C.; Carter, J.D.; Zhang, C.; Gawande, B.; Vrkljan, M.; Janjic, N.; Schneider, D.J. Chemically modified DNA aptamers bind interleukin-6 with high affinity and inhibit signaling by blocking its interaction with interleukin-6 receptor. *J. Biol. Chem.,* **2014**, *289*(12), 8706-8719.
[http://dx.doi.org/10.1074/jbc.M113.532580] [PMID: 24415766]

[49] Lee, C.H.; Lee, S.H.; Kim, J.H.; Noh, Y.H.; Noh, G.J.; Lee, S.W. Pharmacokinetics of a cholesterol-conjugated aptamer against the Hepatitis C Virus (HCV) NS5B Protein. *Mol. Ther. Nucleic Acids,* **2015**, *4*e254
[http://dx.doi.org/10.1038/mtna.2015.30] [PMID: 26440598]

[50] Dougan, H.; Lyster, D.M.; Vo, C.V.; Stafford, A.; Weitz, J.I.; Hobbs, J.B. Extending the lifetime of anticoagulant oligodeoxynucleotide aptamers in blood. *Nucl. Med. Biol.,* **2000**, *27*(3), 289-297.
[http://dx.doi.org/10.1016/S0969-8051(99)00103-1] [PMID: 10832086]

[51] Reinemann, C.; Strehlitz, B. Aptamer-modified nanoparticles and their use in cancer diagnostics and treatment. *Swiss Med. Wkly.,* **2014**, *144*: w13908.
[http://dx.doi.org/10.4414/smw.2014.13908] [PMID: 24395443]

[52] Healy, J.M.; Lewis, S.D.; Kurz, M.; Boomer, R.M.; Thompson, K.M.; Wilson, C.; McCauley, T.G. Pharmacokinetics and biodistribution of novel aptamer compositions. *Pharm. Res.,* **2004**, *21*(12), 2234-2246.
[http://dx.doi.org/10.1007/s11095-004-7676-4] [PMID: 15648255]

[53] Da Pieve, C.; Blackshaw, E.; Missailidis, S.; Perkins, A.C. PEGylation and biodistribution of an anti-MUC1 aptamer in MCF-7 tumor-bearing mice. *Bioconjug. Chem.,* **2012**, *23*(7), 1377-1381.
[http://dx.doi.org/10.1021/bc300128r] [PMID: 22708500]

[54] Pavlova, N.; Kaloudas, D.; Penchovsky, R. Riboswitch distribution, structure, and function in bacteria. *Gene,* **2019**, *708*, 38-48.
[http://dx.doi.org/10.1016/j.gene.2019.05.036] [PMID: 31128223]

[55] Serganov, A.; Nudler, E. A decade of riboswitches. *Cell,* **2013**, *152*(1-2), 17-24.
[http://dx.doi.org/10.1016/j.cell.2012.12.024] [PMID: 23332744]

[56] Aboul-ela, F.; Huang, W.; Abd Elrahman, M.; Boyapati, V.; Li, P. Linking aptamer-ligand binding and expression platform folding in riboswitches: prospects for mechanistic modeling and design. *Wiley Interdiscip. Rev. RNA,* **2015**, *6*(6), 631-650.
[http://dx.doi.org/10.1002/wrna.1300] [PMID: 26361734]

[57] Edwards, A.L.; Batey, R.T. Riboswitches: a common RNA regulatory element. *Nature Education,* **2010**, *3*(9), 9.

[58] Bédard, ASV; Hien, ED; Lafontaine, DA Riboswitch regulation mechanisms: RNA, metabolites and regulatory proteins. *Biochimica et Biophysica Acta (BBA)-Gene Regulatory Mechanisms,* **2020**, 194501.

[59] Findeiß, S.; Etzel, M.; Will, S.; Mörl, M.; Stadler, P.F. Design of artificial riboswitches as biosensors. *Sensors (Basel),* **2017**, *17*(9), 1990.
[http://dx.doi.org/10.3390/s17091990] [PMID: 28867802]

[60] Mehdizadeh Aghdam, E.; Hejazi, M.S.; Barzegar, A. Riboswitches: From living biosensors to novel targets of antibiotics. *Gene,* **2016**, *592*(2), 244-259.
[http://dx.doi.org/10.1016/j.gene.2016.07.035] [PMID: 27432066]

[61] Rekand, I.H.; Brenk, R. Ligand design for riboswitches, an emerging target class for novel antibiotics. *Future Med. Chem.,* **2017**, *9*(14), 1649-1663.
[http://dx.doi.org/10.4155/fmc-2017-0063] [PMID: 28925284]

[62] Seeliger, JC A riboswitch-based inducible gene expression system for mycobacteria. *PLoSOne,* **2012**, *7*(1), e29266.

[63] Wang, Y.X.; Zuo, X.; Wang, J.; Yu, P.; Butcher, S.E. Rapid global structure determination of large RNA and RNA complexes using NMR and small-angle X-ray scattering. *Methods,* **2010**, *52*(2), 180-191.
[http://dx.doi.org/10.1016/j.ymeth.2010.06.009] [PMID: 20554045]

[64] Sakamoto, T. NMR study of aptamers. *Aptamers,* **2010**, *1*, 13-18.

[65] Bains, J.K.; Blechar, J.; de Jesus, V.; Meiser, N.; Zetzsche, H.; Fürtig, B.; Schwalbe, H.; Hengesbach, M. Combined smFRET and NMR analysis of riboswitch structural dynamics. *Methods,* **2019**, *153*, 22-34.
[http://dx.doi.org/10.1016/j.ymeth.2018.10.004] [PMID: 30316819]

[66] Ruigrok, V.J.; Levisson, M.; Hekelaar, J.; Smidt, H.; Dijkstra, B.W.; van der Oost, J. Characterization of aptamer-protein complexes by X-ray crystallography and alternative approaches. *Int. J. Mol. Sci.,* **2012**, *13*(8), 10537-10552.
[http://dx.doi.org/10.3390/ijms130810537] [PMID: 22949878]

[67] Rinnenthal, J.; Buck, J.; Ferner, J.; Wacker, A.; Fürtig, B.; Schwalbe, H. Mapping the landscape of RNA dynamics with NMR spectroscopy. *Acc. Chem. Res.,* **2011**, *44*(12), 1292-1301.

[http://dx.doi.org/10.1021/ar200137d] [PMID: 21894962]

[68] Nomura, Y.; Sugiyama, S.; Sakamoto, T.; Miyakawa, S.; Adachi, H.; Takano, K.; Murakami, S.; Inoue, T.; Mori, Y.; Nakamura, Y.; Matsumura, H. Conformational plasticity of RNA for target recognition as revealed by the 2.15 A crystal structure of a human IgG-aptamer complex. *Nucleic Acids Res.*, **2010**, *38*(21), 7822-7829.
[http://dx.doi.org/10.1093/nar/gkq615] [PMID: 20675355]

[69] Someya, T.; Baba, S.; Fujimoto, M.; Kawai, G.; Kumasaka, T.; Nakamura, K. Crystal structure of Hfq from Bacillus subtilis in complex with SELEX-derived RNA aptamer: insight into RNA-binding properties of bacterial Hfq. *Nucleic Acids Res.*, **2012**, *40*(4), 1856-1867.
[http://dx.doi.org/10.1093/nar/gkr892] [PMID: 22053080]

[70] Haller, A.; Rieder, U.; Aigner, M.; Blanchard, S.C.; Micura, R. Conformational capture of the SAM-II riboswitch. *Nat. Chem. Biol.*, **2011**, *7*(6), 393-400.
[http://dx.doi.org/10.1038/nchembio.562] [PMID: 21532598]

[71] Chen, B.; Zuo, X.; Wang, Y.X.; Dayie, T.K. Multiple conformations of SAM-II riboswitch detected with SAXS and NMR spectroscopy. *Nucleic Acids Res.*, **2012**, *40*(7), 3117-3130.
[http://dx.doi.org/10.1093/nar/gkr1154] [PMID: 22139931]

[72] Al-Hashimi, H.M.; Walter, N.G. RNA dynamics: it is about time. *Curr. Opin. Struct. Biol.*, **2008**, *18*(3), 321-329.
[http://dx.doi.org/10.1016/j.sbi.2008.04.004] [PMID: 18547802]

[73] Ren, A.; Xue, Y.; Peselis, A.; Serganov, A.; Al-Hashimi, H.M.; Patel, D.J. Structural and dynamic basis for low-affinity, high-selectivity binding of L-glutamine by the glutamine riboswitch. *Cell Rep.*, **2015**, *13*(9), 1800-1813.
[http://dx.doi.org/10.1016/j.celrep.2015.10.062] [PMID: 26655897]

[74] Kulik, M.; Mori, T.; Sugita, Y.; Trylska, J. Molecular mechanisms for dynamic regulation of N1 riboswitch by aminoglycosides. *Nucleic Acids Res.*, **2018**, *46*(19), 9960-9970.
[http://dx.doi.org/10.1093/nar/gky833] [PMID: 30239867]

[75] Helm, M.; Kobitski, A.Y.; Nienhaus, G.U. Single-molecule Förster resonance energy transfer studies of RNA structure, dynamics and function. *Biophys. Rev.*, **2009**, *1*(4), 161.
[http://dx.doi.org/10.1007/s12551-009-0018-3] [PMID: 28510027]

[76] Roy, R.; Hohng, S.; Ha, T. A practical guide to single-molecule FRET. *Nat. Methods*, **2008**, *5*(6), 507-516.
[http://dx.doi.org/10.1038/nmeth.1208] [PMID: 18511918]

[77] Buck, J.; Fürtig, B.; Noeske, J.; Wöhnert, J.; Schwalbe, H. Time-resolved NMR methods resolving ligand-induced RNA folding at atomic resolution. *Proc. Natl. Acad. Sci. USA*, **2007**, *104*(40), 15699-15704.
[http://dx.doi.org/10.1073/pnas.0703182104] [PMID: 17895388]

[78] Warhaut, S.; Mertinkus, K.R.; Höllthaler, P.; Fürtig, B.; Heilemann, M.; Hengesbach, M.; Schwalbe, H. Ligand-modulated folding of the full-length adenine riboswitch probed by NMR and single-molecule FRET spectroscopy. *Nucleic Acids Res.*, **2017**, *45*(9), 5512-5522.
[http://dx.doi.org/10.1093/nar/gkx110] [PMID: 28204648]

[79] Bing, T.; Zheng, W.; Zhang, X.; Shen, L.; Liu, X.; Wang, F.; Cui, J.; Cao, Z.; Shangguan, D. Triplex-quadruplex structural scaffold: a new binding structure of aptamer. *Sci. Rep.*, **2017**, *7*(1), 15467.
[http://dx.doi.org/10.1038/s41598-017-15797-5] [PMID: 29133961]

[80] Reinstein, O.; Neves, M.A.; Saad, M.; Boodram, S.N.; Lombardo, S.; Beckham, S.A.; Brouwer, J.; Audette, G.F.; Groves, P.; Wilce, M.C.J.; Johnson, P.E. Engineering a structure switching mechanism into a steroid-binding aptamer and hydrodynamic analysis of the ligand binding mechanism. *Biochemistry*, **2011**, *50*(43), 9368-9376.
[http://dx.doi.org/10.1021/bi201361v] [PMID: 21942676]

[81] Sakamoto, T.; Ennifar, E.; Nakamura, Y. Thermodynamic study of aptamers binding to their target proteins. *Biochimie,* **2018**, *145*, 91-97.
[http://dx.doi.org/10.1016/j.biochi.2017.10.010] [PMID: 29054802]

[82] Martin, J.A.; Mirau, P.A.; Chushak, Y.; Chávez, J.L.; Naik, R.R.; Hagen, J.A.; Kelley-Loughnane, N. Single-round patterned DNA library microarray aptamer lead identification. *J. Anal. Methods Chem.,* **2015**, *2015*: 137489.
[http://dx.doi.org/10.1155/2015/137489] [PMID: 26075138]

[83] Duchardt-Ferner, E.; Gottstein-Schmidtke, S.R.; Weigand, J.E.; Ohlenschläger, O.; Wurm, J-P.; Hammann, C.; Suess, B.; Wöhnert, J. What a Difference an OH Makes: Conformational Dynamics as the Basis for the Ligand Specificity of the Neomycin-Sensing Riboswitch. *Angew. Chem. Int. Ed. Engl.,* **2016**, *55*(4), 1527-1530.
[http://dx.doi.org/10.1002/anie.201507365] [PMID: 26661511]

[84] Ding, J.; Swain, M.; Yu, P.; Stagno, J.R.; Wang, Y-X. Conformational flexibility of adenine riboswitch aptamer in apo and bound states using NMR and an X-ray free electron laser. *J. Biomol. NMR,* **2019**, *73*(8-9), 509-518.
[http://dx.doi.org/10.1007/s10858-019-00278-w] [PMID: 31606878]

[85] Zhao, B.; Guffy, S.L.; Williams, B.; Zhang, Q. An excited state underlies gene regulation of a transcriptional riboswitch. *Nat. Chem. Biol.,* **2017**, *13*(9), 968-974.
[http://dx.doi.org/10.1038/nchembio.2427] [PMID: 28719589]

[86] Krepl, M.; Vögele, J.; Kruse, H.; Duchardt-Ferner, E.; Wöhnert, J.; Sponer, J. An intricate balance of hydrogen bonding, ion atmosphere and dynamics facilitates a seamless uracil to cytosine substitution in the U-turn of the neomycin-sensing riboswitch. *Nucleic Acids Res.,* **2018**, *46*(13), 6528-6543.
[http://dx.doi.org/10.1093/nar/gky490] [PMID: 29893898]

[87] Kim, Y.B.; Wacker, A.; Laer, K.V.; Rogov, V.V.; Suess, B.; Schwalbe, H. Ligand binding to 2′-deoxyguanosine sensing riboswitch in metabolic context. *Nucleic Acids Res.,* **2017**, *45*(9), 5375-5386.
[http://dx.doi.org/10.1093/nar/gkx016] [PMID: 28115631]

[88] Kang, M.; Peterson, R.; Feigon, J. Structural Insights into riboswitch control of the biosynthesis of queuosine, a modified nucleotide found in the anticodon of tRNA. *Mol. Cell,* **2009**, *33*(6), 784-790.
[http://dx.doi.org/10.1016/j.molcel.2009.02.019] [PMID: 19285444]

[89] Kang, M.; Eichhorn, C.D.; Feigon, J. Structural determinants for ligand capture by a class II preQ1 riboswitch. *Proc. Natl. Acad. Sci. USA,* **2014**, *111*(6), E663-E671.
[http://dx.doi.org/10.1073/pnas.1400126111] [PMID: 24469808]

[90] Weickhmann, A.K.; Keller, H.; Wurm, J.P.; Strebitzer, E.; Juen, M.A.; Kremser, J.; Weinberg, Z.; Kreutz, C.; Duchardt-Ferner, E.; Wöhnert, J. The structure of the SAM/SAH-binding riboswitch. *Nucleic Acids Res.,* **2019**, *47*(5), 2654-2665.
[http://dx.doi.org/10.1093/nar/gky1283] [PMID: 30590743]

CHAPTER 4

Applications of NMR Spectroscopy in Medical Diagnosis

Baharudin Ibrahim[*] and **Keshamalini Gopalsamy**

School of Pharmaceutical Sciences, Universiti Sains Malaysia, Penang, Malaysia

Abstract: Nuclear magnetic resonance (NMR) is a special branch of spectroscopy which exploits the magnetic properties of atomic nuclei for molecular elucidation and identification. A technique that was initially developed to analyze chemical and physical molecular structure is now widely used in medical diagnosis. The non-invasiveness, non-destructiveness and simplicity of sample preparation make NMR the preferred technique for metabolomics study. Various body fluids such as urine, saliva, blood, plasma, serum and sweat have been explored to identify potential biomarkers of diseases. Psychiatric disorders, specifically alcohol-use disorder and neurological disorders such as Parkinson's disease, have been investigated with the aid of NMR spectroscopy. Cancer has been one of the most widely studied areas and the research also includes determination of biomarkers which not only could detect the presence of cancer but also potentially predict the various cancer processes in cancer cell lines. Infectious diseases including the compounds produced by the microorganisms such as in tuberculosis and pneumonia have also been explored. Besides, NMR metabolomics has also been used to establish a metabolic fingerprint for risk stratification and early detection of cardiovascular disease (CVD). The samples of subjects with the diseases were collected and the metabolites were compared against controls such as healthy individuals using complex chemometrics and multivariate data analysis such as principal component analysis, partial least square and orthogonal partial least square analyses to distinguish the potential biomarkers. In terms of the various uses of NMR metabolomics in the subject of diagnostic medicine, more improvements to overcome the analytical limitations are expected, making it one of the most notable diagnostic tools of the future. This chapter reviewed some of the published articles in cancer, psychiatric and neurological diseases to provide examples of using NMR spectroscopy in diagnosing human disorders.

Keywords: Cancer, Metabolomics, Neurological disorders, Nuclear Magnetic Resonance Spectroscopy, Psychiatric disorders.

---

[*] **Corresponding author Baharudin Ibrahim:** School of Pharmaceutical Sciences, Universiti Sains Malaysia, Penang, Malaysia; Tel: 604-6535839 / 010-3664181; E-mail: baharudin.ibrahim@usm.my

Atta-ur-Rahman and M. Iqbal Choudhary (Eds.)
All rights reserved-© 2020 Bentham Science Publishers

INTRODUCTION

Since ancient times, humans have utilized urine, saliva and other bodily fluids for the identification of various ailments. The advancement and utilization of analytical techniques for the evaluation of these biofluids have brought about the discovery of various disease biomarkers [1]. The integration of NMR, mass spectrometry (MS) and multivariate statistical techniques became the cornerstone for metabolomics-based disease diagnosis [2]. Metabolomics can be described as an in-depth study of chemical processes involving metabolites in a biological system [3]. Another terminology which is frequently used interchangeably with metabolomics is metabonomics. Metabonomics is defined as the quantitative measurement of metabolic responses of living systems against time to pathophysiological stimuli or genetic modification [4]. The primary objective of NMR-based medical diagnosis is to identify metabolites that precisely correspond to a particular disease for the early detection and treatment of said illness. In this chapter, we will analyze recent publications and highlight the advancements in experimental techniques, sample preparation, discovery and quantification of metabolites using chemometric tools used to identify biomarkers through NMR. One study for each disease will be reviewed in detail to explain this technique.

WORKING PRINCIPLE OF NMR IN MEDICAL DIAGNOSIS

The principle behind NMR is that the nuclei in atoms are charged and hence is detectable by NMR, due to the formation of magnetic dipoles. When an external magnetic field is applied through NMR spectrometer, the base energy is shifted to a higher energy level. The energy transfer produces a wavelength that is measured and processed to produce NMR spectrum for the particular nucleus. With the help of chemometrics software, the area under the curve or peak height/intensity of the spectra can be calculated and used to identify significant/important compounds of diseases. This approach is known as metabolomics. The two main system used in metabolomics study are Nuclear magnetic resonance (NMR) spectroscopy and mass spectrometry (MS). They both have different advantages and setbacks but both techniques are able to provide complementary information. The MS is usually combined with liquid chromatography (LC) or gas chromatography (GC) and has higher sensitivity compared to NMR. Thus, it is the preferred choice for metabolomics studies where a particular group of compounds are being targeted such as lipid compounds [5]. Nevertheless, MS sample preparation is extensive. It usually involves many steps such as solvent extraction, ultrafiltration, solid-phase extraction and a chemical derivatization. The presence of other chemical species may influence matrix effects, ionization suppression and enhancement and cause inconsistent results. Furthermore, samples are destroyed in the process [6, 7].

Meanwhile, NMR is fast and does not require tedious sample processing [8]. The samples are also preserved thus can be stored and re-run for further analysis. Newer NMR machines allow automation and can run samples in large quantities [9]. It is also non-selective thus is a preferred choice for bulk analysis in metabolomics studies to identify discriminating metabolites without any prior knowledge [10]. In addition to that, NMR provides insight to the molecular dynamics and mobility of a particular metabolite [11]. One major drawback of NMR is its sensitivity and resolution which is lower compared to MS [12]. However, newer NMR is continuously being developed with higher sensitivity and resolution.

Besides *in-vitro* NMR spectroscopy, *in-vivo* magnetic resonance spectroscopy is also a non-invasive method which can complement the magnetic resonance imaging (MRI) in the characterization of tissue and can be used to study metabolic changes in brain related disorders such as stroke, depression, tumors, dementia and seizures. In comparison, traditional investigational techniques such as biopsy is more invasive and has more risks and side effects. In this chapter, selected studies of NMR-based metabolomics applications with reference to specific diseases are discussed.

NMR in the Diagnosis of Lung Cancer

Background of Lung Cancer

Lung cancer poses a serious health burden in most developed nations and it is estimated that there will be roughly 228,150 new cases of lung and bronchus cancer in 2019 [13]. Recent studies suggest that there may be a strong association between inherited genes and development of lung cancer. Nevertheless, there are very few genes that have been associated to lung cancer hitherto [14]. Besides hereditary factor, smoking has been shown to be one of the main risk factors for lung cancer [15]. Similar to other cancers, pathogenesis of lung cancer is induced through carcinogens, followed by a period of promotion and progression in a multistep process [16, 17]. Even though cancer risk may decrease after smoking cessation, another carcinogen may still carry on the process [18]. Table **1** shows the classification of lung cancers and its features.

Early detection is imperative to increase patient survival in lung cancer; nonetheless, available diagnostic techniques are insufficient. Diagnostic work-up for lung cancer still relies heavily on clinical perspectives and no single clinically based algorithm can be applied to all the cases [1, 19]. Definitive diagnosis of lung cancer is primarily based on the histopathological analysis of the lung cells. Due to the lack of screening tests and the onset of tumor growth generally do not display any signs or symptoms, diagnosis is frequently deferred. Therefore, to increase the survival rate of patients with lung cancer, there is a significant need for the development of methods which allows for an earlier detection of lung cancer [20]. To address this issue, Carrola *et al.* utilized NMR fingerprinting of human urine to identify metabolic signatures of lung cancer, which may be of use in assisting diagnosis and in providing key insights on lung cancer metabolism [21]. Another study conducted by Rocha and colleagues have elucidated metabolites that can differentiate between adenocarcinoma and squamous cell carcinoma through NMR fingerprinting [22].

Table 1. Classification of invasive lung cancer. Adapted from Zheng, 2016 & Lewis *et al.*, 2014 [23, 24].

Lung Cancer Type	Location in the Lung	Features
Adenocarcinoma	Peripheral	• Accounts for more than 40% of all lung cancer cases • More common in non-smokers • Characterized by malignant neoplasm with glandular differentiation, pneumocyte phenotype or mucin production
Squamous cell carcinoma	Central & peripheral	• Accounts for 20% of all lung cancer • Significant association with cigarette smoking • Arises from large airway epithelial cells
Small-cell lung carcinoma (SCLC)	Central	• Strongest smoking association • Rapid growth and early distant metastasis • This results in worst prognosis
Large cell carcinoma	Peripheral	• Somewhat similar to adenocarcinomas • Normally located peripherally, seems bulky and necrotic in appearance

Study by Carrola et al. (2011)

Sample Collection

High-resolution ^{1}H-NMR were used to analyze urine samples from lung cancer patients (n = 71) and healthy controls (n = 54), and their spectral profiles were subjected to multivariate analysis. In the control group, 19 of them were smokers, 9 ex-smokers, and 25 non-smokers. The smoking status of one control was unknown. Sample of urine was collected from each individual in the morning in a sterile container. Then, fractions of about1 mL were transferred into sterile cryovials. These were stored at -80 °C.

Sample Preparation

The samples were left in room temperature to thaw them prior to analysis. They were then centrifuged at 8000 rpm for 5 min to remove any particles. A stock buffer solution of KH2PO4v1.5 M in D_2O was prepared containing 0.1% TSP-*d*4 and 2mM sodium azide (NaN_3). TSP is used as a chemical shift reference and NaN_3 is a bacteriostatic agent. About 60 μL of the buffer solution was added to 540 μL of urine and vortexed. The pH was adjusted to 7.00 and 550 μL of the sample was transferred to a 5 mm NMR tube.

Main Findings

This study successfully discriminated between healthy subjects and those with lung cancer. Fig. (**1**) shows the ^{1}H NMR spectra of urine from a healthy subject and a lung cancer patient. Creatinine, trimethylamine-N-oxide/betaine, hippurate, citrate, R-ketoglutarate, and glycine gave out the most intense signals. Using Monte Carlo Cross Validation (MCCV), the classification model resulted in 93% sensitivity, 94% specificity and an overall classification rate of 93.5%. A *p*-value < 0.01 was considered statistically significant. Fig. (**2**) shows the disparity between the amount of metabolites in control and cancer samples. The following metabolites produced the strongest signals: α-ketoglutarate, glycine, creatinine, TMAO/betaine, hippurate and citrate. Both hippurate and trigonelline were reduced in cancer patients whereas β-hydroxyisovalerate, α-hydroxyisobutyrate, N-acetylglutamine, and creatinine were higher in cancer patients. The metabolites trigonelline, hippurate and formate in urine showed strong positive correlation with baseline lung function in adults with and without COPD [25]. Hippurate is an organic compound often found in urine and is the glycine conjugate of benzoic acid. Increased consumption of food high in phenol has shown to raise hippurate levels in the body [26]. These results reflect those of Cheng *et al.* who also found that hippurate level was reduced in subjects with colorectal cancer compared to healthy subjects. Trigonelline is an alkaloid which is present in considerable amounts in coffee [27]. The present study found this compound to be reduced in patients with lung cancer. Trigonelline has been identified as a biomarker of coffee consumption [28]. Chen *et al.*, (2009) also found reduced trigonelline in patient with liver cancer [29].

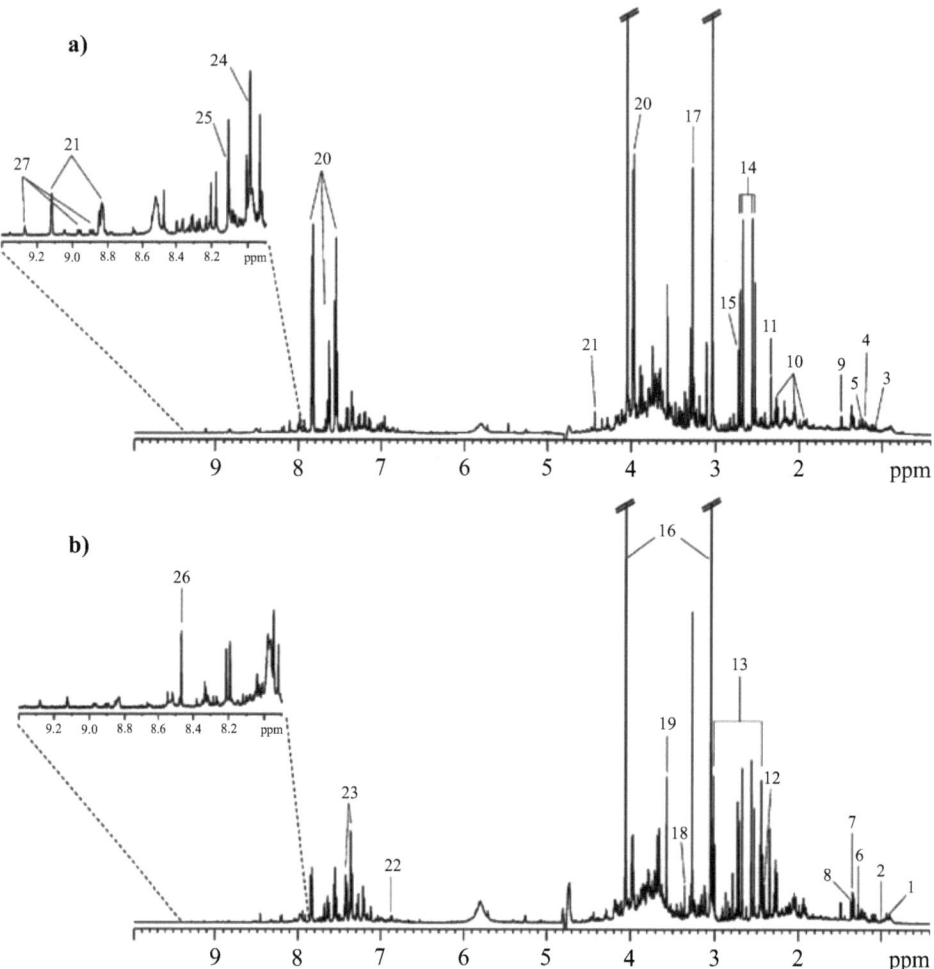

Fig. (1). The 500 MHz ¹H NMR spectra of urine from **(a)** a healthy (control) subject, and **(b)** a lung cancer patient. Signal assignment: 1, R-hydroxybutyrate; 2, valine; 3, isobutyrate; 4, -aminoisobutyrate; 5, methyl--hydroxybutyrate; 6, -hydroxyisovalerate; 7, lactic acid and threonine; 8, R-hydroxyisobutyrate; 9, alanine; 10, N-acetylglutamine; 11, pyruvate; 12, succinate; 13, R-ketoglutarate; 14, citrate; 15, dimethylamine; 16, creatinine; 17, trimethylamine-N-oxide and betaine; 18, scyllo-inositol; 19, glycine; 20, hippurate; 21, trigonelline; 22, p-hydroxyphenylacetate; 23, phenylacetylglycine; 24, histidine; 25, 3-methylhistidine; 26, formate; 27, trigonellinamide. Adapted from Carrola *et al.*, 2011 [21].

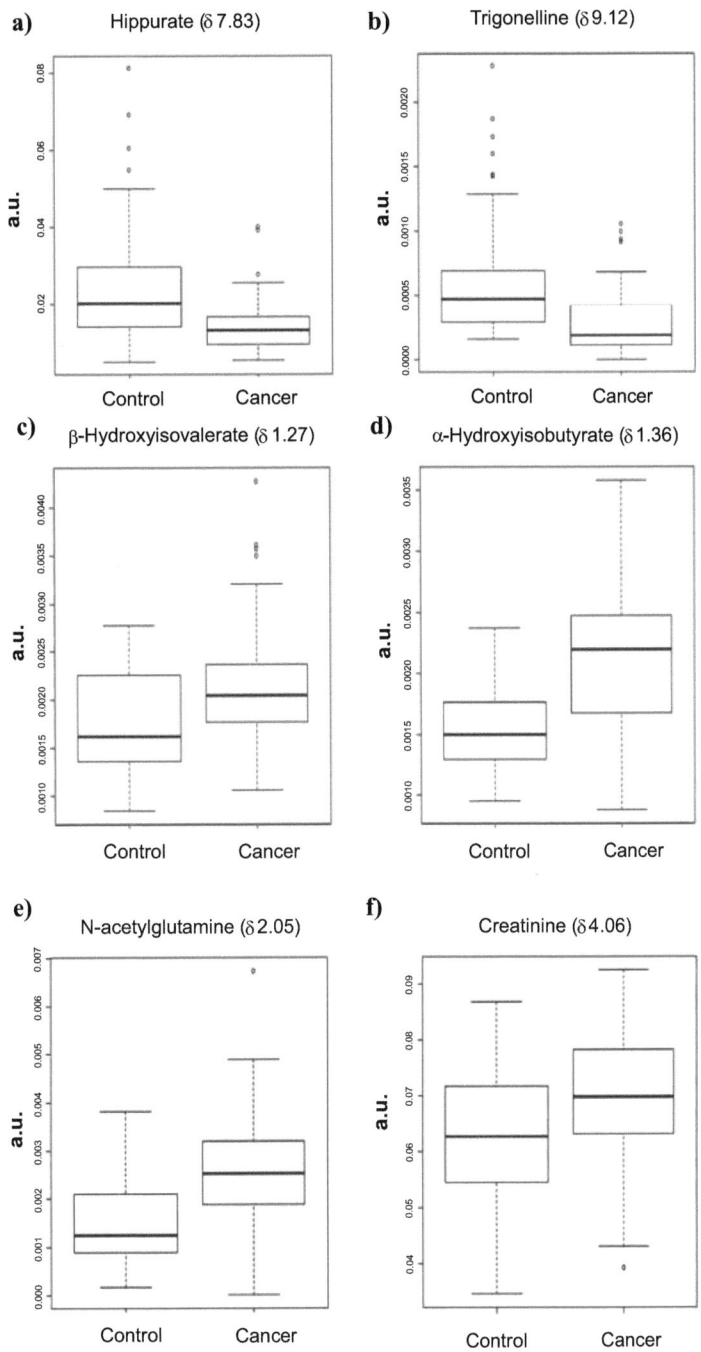

Fig. (2). Boxplots representing the median, 1st and 3rd quartiles (Q1 and Q3), minimum and maximum of integral values of selected signals (normalized to the total spectral area) for healthy controls (n) 54 and lung cancer patients (n) 71: **(a)** hippurate **(b)** trigonelline **(c)** β-hydroxyisovalerate **(d)** α-hydroxyisobutyrate **(e)** N-acetylglutamine **(f)** creatinine. Adapted from Carrola *et al.*, 2011 [21].

Given the fact that gender and age influences the metabolic composition of urine, these confounding factors have been modeled in this study. Test for gender variation resulted in negative Q^2 values, indicating poor predictive ability. Besides, using age as a classifier, the PLS-DA model demonstrated poor predictive power (negative Q^2), a low classification rate, specificity, and sensitivity. However, a good separation was obtained when the presence of the disease was used as the classifier for the PLS-DA modeling. The overall classification rate was 84%. To investigate if smoking habits of the subjects could be a potential confounding factor, control subjects were grouped according to their smoking habits and modeled by PLS-DA, along with the cancer class., the specificity. A decrease to 81% in specificity was observed when only smokers were included in the control group. The overall classification rate and sensitivity remained high. On the other hand, when only non-smokers were included in the control group, class discrimination did not improve whereby Q2 was lower than 0.5 and the classification rate decreased to 90%. Based on these findings, we can say that smoking may not have high impact on the model development. The key discriminating biomarkers/metabolites were hippurate and trigonelline, β-hydroxyisovalerate, α-hydroxyisobutyrate, N-acetylglutamine, and creatinine (Table 2). This study provided proof-of-concept of the useful potential of NMR-based metabonomics in identifying biomarkers for lung cancer in urine. Future studies can work on determining the specificity of these markers for lung cancer.

Table 2. Metabolites involved in the discrimination between cancer and healthy subjects Carrola *et al.*, 2011 [21].

Metabolite	δ 1H (ppm)	p-value
Hippurate	7.83	4.59×10^{-6}
Trigonelline	9.12	3.47×10^{-7}
Trigonellinamide	9.28	0.0323
β-hydroxyisovalerate	1.27	9.78×10^{-4}
α-hydroxyisobutyrate	1.36	2.54×10^{-11}
acetylglutamine	2.05	1.44×10^{-7}
Creatinine	4.06	4.88×10^{-4}
Citrate	2.54	0.0213
Phenylacetylglycine	7.37	0.263

[a]Average changes determined by spectral integration of the signal underlined in the second column. P-value determined by comparison of mean integrals between the two groups.

Limitations

Nevertheless, it is important to be cognizant of the limitations and possible confounding factors that have a bearing on the metabolomics profile of subjects. In this study, the three potential confounding factors are age, gender and smoking habits. It is now well established from a number of studies that intrinsic factors (such as age and gender) and extrinsic factors (such as diet, smoking and exercise) regulate the metabolic composition of human urine, resulting in high inter-individual variability [30 - 33]. Hence, the accuracy of the results could have been compromised since the subjects in the control group were significantly younger compared to the disease group. Nevertheless, these limitations have been addressed as explained in the main findings.

Summary

This study is first of its kind that has identified markers of lung cancer in urine. It has demonstrated that ^{1}H NMR spectroscopy along with statistical analysis has identified biomarkers in the urine samples of individuals diagnosed with lung cancer, which allows for the discrimination from healthy subjects with high sensitivity and specificity. Factors such as age, gender and smoking habits were not found to have a huge impact on this discrimination. However, future studies should consider other confounding factors such as diet and this should be assessed in the upcoming studies.

NMR in the Diagnosis of Alcohol Use Disorder (AUD)

Background of Alcohol Use Disorder (AUD)

Addiction to alcohol is one of the most alarming and persistent health issues globally. Alcoholism results from an impaired control over alcohol intake and it is often hard to combat alcohol abuse without appropriate support and guidance. The primary method which is employed to diagnose alcohol use disorders in clinical setting are questionnaires. For instance, the Cut Down, Annoyed, Guilty and Eye Opener (CAGE) questionnaire [34], Alcohol Use Disorders Identification Test (AUDIT) and Michigan Alcoholism Screening Test (MAST) comprise of a series of questions that are used to differentiate between acceptable and harmful alcohol intake [35].

Nonetheless, these questionnaires lack in reliability and are relatively brief. The vast majority of AUD subjects are hesitant to reveal their alcohol consumption. Hence, NMR-based metabolomics have emerged as powerful platforms to distinguish new biomarkers to differentiate alcohol-dependent, non-AD alcohol drinkers and controls using metabolomics [36]. Hitherto, not many metabolomics studies have been carried out to investigate the impact of alcohol intake on human metabolome.

Recent studies conducted by Mostafa *et al.* [36, 37], discovered novel metabolic fingerprints in biofluids that can discern AUD individuals from social drinkers and alcohol naive subjects. NMR-based metabolomics approach was employed to study urine and plasma samples to identify discriminatory biomarkers that can be used to diagnose AUD.

Study by Mostafa et al. (2016, 2017)

Sample Population

In order to ensure the identified biomarkers were for alcohol-dependence (AD), this study included two control groups *i.e.* social drinkers and alcohol-naïve. The AD were newly diagnosed patients based on the AD questionnaire and/or the National Institutes of Health (NIH) criteria of AD and without any treatment started [38]. Alcohol-dependent is defined as those who drinks at least 15 drinks per week or who has at least 5 drinks per occasion in a week. The social drinkers are those who drink alcohol occasionally but did not fulfill the NIH AD criteria. This was ascertained by the physician. Alcohol naïve is those who has never drank alcohol. Exclusion criteria is those who had been diagnosed with HIV or any blood transmitted disease, liver cirrhosis, or acute or chronic infectious diseases.

Sample Preparation, Metabolites Determination and Statistical Analysis

About 5 mL of blood plasma was centrifuged at 12000 rpm for 10 min at 4 °C to separate the plasma and 10 mL of urine were collected and stored at -80°C. The samples were analyzed using Nuclear Magnetic Resonance (NMR) spectroscopy (BRUKERAscendTM500, Germany). The NMR spectra were aligned and normalized to TSP and binned to 0.04ppm using TopSpin 3.2 (BRUKER BioSpin, Germany) and AMIX (BRUKER BioSpin, Germany). The ethanol and water peaks were excluded and the data was further imported to Soft Independent Modelling of Class Analogies (SIMCA 13.0.3, Umetrics) and principal component analysis (PCA) and orthogonal partial least square-discriminant analysis (OPLS-DA) were done. All the imported data were mean-centered and Pareto scaled before the multivariate analysis. Further univariate and multivariate logistic regression were performed using Statistical Package for the Social Sciences (SPSS 21 (IBM, USA)software) to develop a discriminating model between the AD group, non-AD social drinkers and controls of AUD The identity of the metabolites were determined using metabolomics databases *i.e.* Bruker Biofluid Reference Compound Database (B-BIOREFCODE), the Chenomx, the

Human Metabolome Database (HMDB) and the Biological Magnetic Resonance Data Bank (BMRB-metabolomics). To ratify the identity of the metabolites, 2D Heteronuclear Single Quantum Coherence (HSQC) experiments were carried out.

Main Findings

1, 2-propanediol, alanine, cis-aconitic acid, citric acid, 2-hydroxyisovaleric acid and lactic acid were identified as biomarkers of AUD in urine samples with high specificity and accuracy (Fig. **3**). Except for 2-hydroxyisovaleric acid, all the other biomarkers appear naturally in urine but at much lower concentrations. It has been observed that after persistent intake of alcohol, the concentrations of these biomarkers were greatly increased. The presence of some of these metabolites can be explained by looking at the biological pathways. For example, lactic acid is produced by the metabolism of ethanol. This could be due to the increased NADH/NAD+ ratio arising from ethanol metabolism which cause the conversion of pyruvate to lactate instead of glucose [39]. Landaas *et al.* also has reported the occurrence of 2-hydroxyisovaleric acid in patients with lactic acidosis [40]. Meanwhile, citric acid and cis-aconitic acid are intermediates of citric acid (Krebs) cycle and has been reported to increase in cases of lactic acidosis too [41].

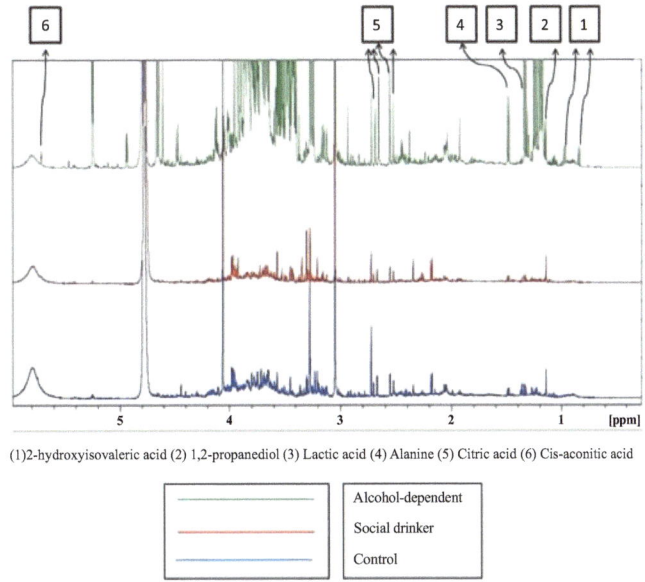

(1) 2-hydroxyisovaleric acid (2) 1,2-propanediol (3) Lactic acid (4) Alanine (5) Citric acid (6) Cis-aconitic acid

Alcohol-dependent
Social drinker
Control

Fig. (3). Example of one full ^{1}H-NMR spectrum each of the three groups. The numbers in the box on the top of the selected peaks indicate the biomarkers in urine that are different between the three groups based on the peak height/intensity. AUD (green): Compounds (2), (3), (4), (5) and (6) are higher in this group compared to other groups. Social drinkers (red) and control (blue): Compound (1) is higher compared to AUD. Adapted from Mostafa *et al.*, 2016 [36].

In plasma, two biomarkers have been identified namely acetic acid and propionic acid. Acetic acid is the final product of ethanol oxidation and is an endogenous metabolite. However, further exploration of the spectra of a male social drinker has confirmed that the discriminatory ability of this metabolite was not compromised by recent non-chronic drinking. As explained above, ethanol metabolism increased NADH/NAD+ ratio and lead to increased lactate formation which will cause the accumulation of propionate [42]. In Fig. (**4**), spectra of propionic acid were compared between AUD, social drinker and a control. The absence of this metabolite peaks in the NMR spectra of social drinker and control proved that it is a reliable biomarker for AUD.

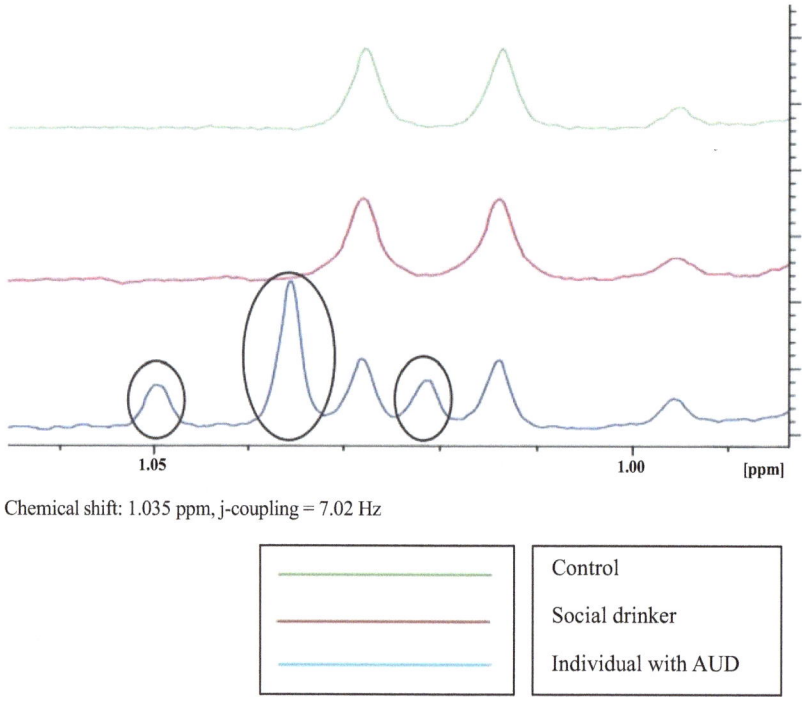

Fig. (4). ^{1}H-NMR spectra of the three groups showing propionic acid (triplet) in individual with AUD. AUD (blue): Propionate formation increases with chronic alcohol drinking. This peak is absent in social drinkers (red) and the control group (green). The 1H-NMR spectra of a male social drinker was checked to ensure that propionic acid is a true biomarker for AUD and not because of recent consumption of alcohol. Adapted from Mostafa *et al.*, 2017 [37].

Summary

Both the urine and plasma metabolites studies emphasize the notion that ^{1}H NMR is an effective platform in metabolomics to identify discriminating biomarkers in urine and plasma with for AUD. Further confirmation with pathway comparison and analysis provide further prove that the identified biomarkers could be used for

AUD determination. High specificity and accuracy of these biomarkers could enable them to be used in clinical settings in future.

NMR in the Diagnosis of Parkinson's Disease

Background of Parkinson's Disease

Parkinson's disease (PD) is only second to Alzheimer's disease (AD) in terms of most common neurodegenerative disorder [43]. It is a neurodegenerative disorder caused by the degeneration of dopamine-generating cells, resulting in shaking, rigidity, reductions in movement and difficulties in walking [44]. The cause of this cell death that is characteristic of PD is still unknown. Despite various studies that have been carried out, the root cause of PD is yet to be established [45]. It is widely accepted that the combined influence of genetic susceptibility, environmental exposure and complex genetic environmental interactions contributed to PD. In addition to that, aging has also been identified as an important risk factor in PD [46, 47]. Mechanisms such as oxidative stress, apoptosis, excitotoxicity and inflammatory responses have been proposed to describe cell death in PD [48 - 50].

Similar to other neurological disorders, the diagnosis of Parkinson's disease (PD) is still a challenge, especially at the onset of the disease [51]. There are no specific tests available such as brain scan to determine if an individual has Parkinson's disease. The current practice is to analyze their medical history and conduct an in-depth neurological examination, looking for the presence of two or more cardinal signs to be present. Rigidity, rest tremor, bradykinesia, and decline in postural reflexes are the main cardinal signs of PD [52]. ^{1}H NMR is the prevailing noninvasive method that has been utilized to investigate central nervous system pathologies.

Study by Ahmed et al. (2009)

Sample Collection

All samples were obtained from the Department of Neurology at SRM Hospital, Tamil Nadu, India. A total of 80 samples were obtained, with 43 PD patients who were free of drugs and the remaining were samples of healthy controls. Both groups were matched for age and gender. All 43 patients with PD were assessed on family history of PD, date of the symptom onset and date of diagnosis. The disease status was evaluated by neurological specialists from hospital and the disease stages were classified according to Unified Parkinson's Disease Rating Scale (UPDRS).

Sample Preparation

About 4 ml of blood samples were collected in EDTA tubes (Becton Dickinson, Franklin Lakes, NJ). Eppendorf centrifuge was used to centrifuge the samples for 5 minutes at 14,000 rpm. The resultant plasma portion was shifted to new Eppendorf tubes and was kept at -20°C before further processing. For the extraction of metabolites from the sample, Nanosep 3KD (Pall Co., New York, USA) micro centrifuge tube was used to process the plasma. The metabolite extracts were obtained and dissolved in D_2O (Merck KGaA, Darmstadt, Germany). Next, 500 µM of 2,2-dimethyl-2-silapentane-5- sulfonate (DSS) (Sigma Chemical Co., St. Louis, MO, U.S.A) were added as an internal standard in preparation for ^{1}H-NMR analysis.

Metabolites Determination and Statistical Analysis

Partial least square discriminant analysis showed a marked separation between control subjects (n = 37) and PD patients (n = 43) (Fig. **5**). Chenomx NMR 5.1 software (Chenomx. Inc., Edmonton, Canada) was used to obtain discriminating metabolites between the two groups. 22 metabolites in the plasma was determined and quantified by comparing them with a library of 292 metabolites of ^{1}H-NMR spectra. The calibration of chemical shift was made by DSS internal standard (0.0 ppm) resonance. Out of the 22 metabolites, the concentrations of ethymalonate, pyruvate, myoinositol, sorbitol and propylene glycol was higher in the patient samples. The average difference in concentrations of metabolites between normal and patients was showed in heat map (Fig. **6**).

Main Findings

The study focused on 22 metabolites that have been shown to play a role in PD. These 22 metabolites are critical in mitochondrial function and other associated pathways. The key metabolites identified in this study are acetate, citrate, pyruvate, succinate and malate. These compounds vary significantly in PD plasma samples and contribute to the major differentiation of PD from normal samples in PLS-DA analysis. A reduction was observed in acetate, citrate, succinate and malate, while pyruvate concentration was increased in PD samples. Altered pyruvate metabolism has been documented in several neurodegenerative disorders, including Alzheimer's disease (AD) and Leigh syndrome [54]. In addition to that, studies have also found that this could be attributed to reduced activity of the pyruvate dehydrogenase complex (PDHC) in affected areas of brains of patients with neurological diseases [55].

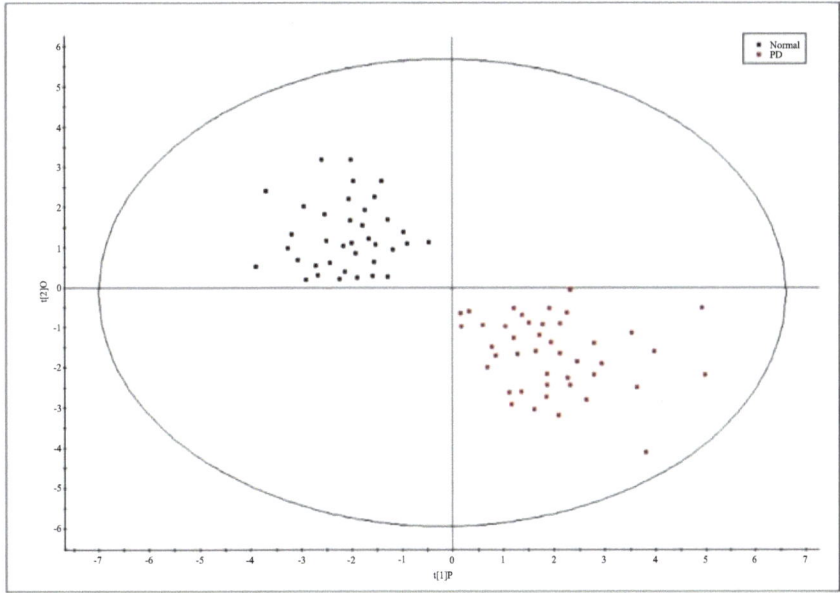

Fig. (5). Partial least square discriminant analysis. PLS-DA scores plot showing a significant separation between control subjects (n = 37) and PD patients (n = 43). Black square = controls; red square = PD patients. Adapted from Ahmed *et al.*, 2009 [53].

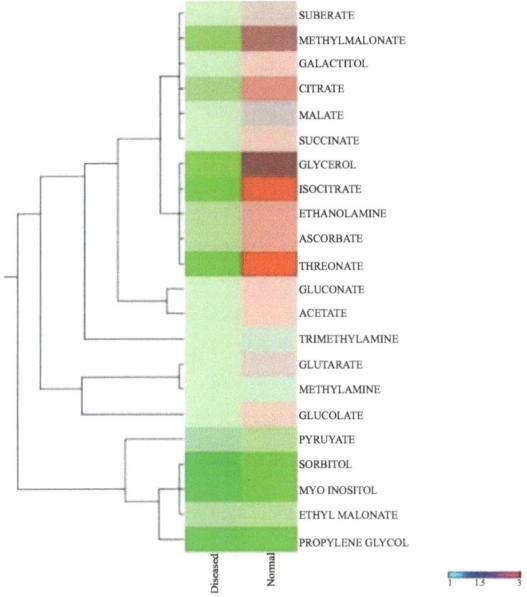

Fig. (6). Heat map differentiation of metabolite. Average metabolite variability of blood plasma between PD patients (n = 43) and healthy controls (n = 37) are shown. The heat map depicts high (red) and low (green) relative levels of metabolite variation. Adapted from Ahmed *et al.*, 2009 [53].

All of the studies mentioned above suggest that there could be a strong link between aberrant pyruvate metabolism and neurodegenerative disorders. Therefore, understanding the metabolism pathway of pyruvate and other metabolites involved in this pathway may provide a potential candidate biomarker in the diagnosis of PD.

Limitations

They only targeted 22 metabolites (assuming these cause PD) hence the study is restricted to these metabolites only.

NMR in the Diagnosis of Other Diseases

Nuclear magnetic resonance metabolomics has also been used in other diseases such as in infectious and cardiovascular diseases. Zhou *et al.*, studied blood samples from 38 active TB patients and 39 healthy controls [56]. The samples were analyzed using NMR spectroscopy. The study found seventeen metabolites significantly different between the two groups. The results obtained from this study revealed up-regulation of acetone, tyrosine, 1-methylhistidine, acetoacetate, glutamate, glutamine, isoleucine, lactate, lysine, nicotinate, phenylalanine and pyruvate, and down-regulation of glycerolphosphocholine, alanine, formate, glycine, and low-density lipoproteins (LDL) in TB patients relative to healthy controls. Vignoli *et al.*, investigated the metabolomic fingerprint of 978 acute myocardial infarction patients through NMR spectroscopy. In this study, they showed the ability of this technique to stratify those who are at a higher risk of fatality and hence, might allow physicians to aim towards precision medicine [57].

CONCLUSION

The NMR spectroscopy was designed to elucidate physical, chemical, electronic and structural information about molecules and has evolved from being used in analytical chemistry, molecular physics and biochemistry to the medical field. The machine itself has improved tremendously in terms of sensitivity and reproducibility and this has opened its' wider use in medical diagnosis. With the potential of identifying biomarkers of diseases and drug response, the future of NMR spectroscopy in medical application is positive. Establishment of phenome centers in several countries which uses NMR to analyze serum, plasma and urine samples to study human health is a testament to the benefits of NMR in medical diagnosis.

CONSENT FOR PUBLICATION

Not applicable.

CONFLICT OF INTEREST

The authors confirm that this chapter contents have no conflict of interest.

ACKNOWLEDGEMENTS

Declared none.

REFERENCES

[1]   Gebregiworgis, T.; Powers, R. Application of NMR metabolomics to search for human disease biomarkers. *Comb. Chem. High Throughput Screen.,* **2012**, *15*(8), 595-610.
[http://dx.doi.org/10.2174/138620712802650522] [PMID: 22480238]

[2]   Marshall, D.D.; Powers, R. Beyond the paradigm: Combining mass spectrometry and nuclear magnetic resonance for metabolomics. *Prog. Nucl. Magn. Reson. Spectrosc.,* **2017**, *100*, 1-16.
[http://dx.doi.org/10.1016/j.pnmrs.2017.01.001] [PMID: 28552170]

[3]   Clarke, C.J.; Haselden, J.N. Metabolic profiling as a tool for understanding mechanisms of toxicity. *Toxicol. Pathol.,* **2008**, *36*(1), 140-147.
[http://dx.doi.org/10.1177/0192623307310947] [PMID: 18337232]

[4]   Lindon, J.C.; Nicholson, J.K.; Holmes, E. *The Handbook of Metabonomics and Metabolomics*; Elsevier: London, **2011**.

[5]   Elipe, M.V.S. Advantages and disadvantages of nuclear magnetic resonance spectroscopy as a hyphenated technique. *Anal. Chim. Acta,* **2003**, *497*(1-2), 1-25.
[http://dx.doi.org/10.1016/j.aca.2003.08.048]

[6]   Causon, T.J.; Hann, S. Review of sample preparation strategies for MS-based metabolomic studies in industrial biotechnology. *Anal. Chim. Acta,* **2016**, *938*, 18-32.
[http://dx.doi.org/10.1016/j.aca.2016.07.033] [PMID: 27619083]

[7]   Aretz, I.; Meierhofer, D. Advantages and pitfalls of mass spectrometry based metabolome profiling in systems biology. *Int. J. Mol. Sci.,* **2016**, *17*(5), 632.
[http://dx.doi.org/10.3390/ijms17050632] [PMID: 27128910]

[8]   Dunn, W.B.; Bailey, N.J.; Johnson, H.E. Measuring the metabolome: current analytical technologies. *Analyst (Lond.),* **2005**, *130*(5), 606-625.
[http://dx.doi.org/10.1039/b418288j] [PMID: 15852128]

[9]   Nicholson, J.K.; Lindon, J.C. Systems biology: Metabonomics. *Nature,* **2008**, *455*(7216), 1054-1056.
[http://dx.doi.org/10.1038/4551054a] [PMID: 18948945]

[10] Alonso, A.; Marsal, S.; Julià, A. Analytical methods in untargeted metabolomics: state of the art in 2015. *Front. Bioeng. Biotechnol.,* **2015**, *3*, 23.
[http://dx.doi.org/10.3389/fbioe.2015.00023] [PMID: 25798438]

[11] Capati, A; Ijare, OB; Bezabeh, T Diagnostic Applications of Nuclear Magnetic Resonance–Based Urinary Metabolomics. *Magn. Reson. Insights.,* **2017**, *10*, 1178623X17694346.
[http://dx.doi.org/10.1177/1178623X17694346]

[12] Nagana Gowda, G.A.; Raftery, D. Can NMR solve some significant challenges in metabolomics? *J. Magn. Reson.,* **2015**, *260*, 144-160.
[http://dx.doi.org/10.1016/j.jmr.2015.07.014] [PMID: 26476597]

[13] SEER, N. *Cancer Stat Facts: Lung and Bronchus Cancer,* **2016**.

[14] Koeller, D.R.; Chen, R.; Oxnard, G.R. Hereditary Lung Cancer Risk: Recent Discoveries and Implications for Genetic Counseling and Testing. *Curr. Genet. Med. Rep.,* **2018**, *6*(2), 83-88.
[http://dx.doi.org/10.1007/s40142-018-0140-2]

[15] Hecht, S.S. Tobacco smoke carcinogens and lung cancer. *J. Natl. Cancer Inst.,* **1999**, *91*(14), 1194-1210.
[http://dx.doi.org/10.1093/jnci/91.14.1194] [PMID: 10413421]

[16] Cooper, G.M.; Hausman, R.E. The development and causes of cancer. *The cell: A Molecular Approach.,* **2000**, 725-66.

[17] Zöchbauer-Müller, S.; Gazdar, A.F.; Minna, J.D. Molecular pathogenesis of lung cancer. *Annu. Rev. Physiol.,* **2002**, *64*(1), 681-708.
[http://dx.doi.org/10.1146/annurev.physiol.64.081501.155828] [PMID: 11826285]

[18] MacKinnon, A.C.; Kopatz, J.; Sethi, T. The molecular and cellular biology of lung cancer: identifying novel therapeutic strategies. *Br. Med. Bull.,* **2010**, *95*, 47-61.
[http://dx.doi.org/10.1093/bmb/ldq023] [PMID: 20643690]

[19] Ganguly, D. Lung cancer-diagnostic problems. *J. Indian Acad. Clin. Med.,* **2012**, *13*, 2.

[20] Midthun, D.E. Early detection of lung cancer. *F1000 Res.,* **2016**, *5*, 5.
[http://dx.doi.org/10.12688/f1000research.7313.1] [PMID: 27158468]

[21] Carrola, J.; Rocha, C.M.; Barros, A.S.; Gil, A.M.; Goodfellow, B.J.; Carreira, I.M.; Bernardo, J.; Gomes, A.; Sousa, V.; Carvalho, L.; Duarte, I.F. Metabolic signatures of lung cancer in biofluids: NMR-based metabonomics of urine. *J. Proteome Res.,* **2011**, *10*(1), 221-230.
[http://dx.doi.org/10.1021/pr100899x] [PMID: 21058631]

[22] Rocha, C.M.; Barros, A.S.; Goodfellow, B.J.; Carreira, I.M.; Gomes, A.; Sousa, V.; Bernardo, J.; Carvalho, L.; Gil, A.M.; Duarte, I.F. NMR metabolomics of human lung tumours reveals distinct metabolic signatures for adenocarcinoma and squamous cell carcinoma. *Carcinogenesis,* **2015**, *36*(1), 68-75.
[http://dx.doi.org/10.1093/carcin/bgu226] [PMID: 25368033]

[23] Zheng, M. Classification and pathology of lung cancer. *Surg. Oncol. Clin. N. Am.,* **2016**, *25*(3), 447-468.
[http://dx.doi.org/10.1016/j.soc.2016.02.003] [PMID: 27261908]

[24] Lewis, D.R.; Check, D.P.; Caporaso, N.E.; Travis, W.D.; Devesa, S.S. US lung cancer trends by histologic type. *Cancer,* **2014**, *120*(18), 2883-2892.
[http://dx.doi.org/10.1002/cncr.28749] [PMID: 25113306]

[25] McClay, J.L.; Adkins, D.E.; Isern, N.G.; O'Connell, T.M.; Wooten, J.B.; Zedler, B.K.; Dasika, M.S.; Webb, B.T.; Webb-Robertson, B.J.; Pounds, J.G.; Murrelle, E.L.; Leppert, M.F.; van den Oord, E.J. (1)H nuclear magnetic resonance metabolomics analysis identifies novel urinary biomarkers for lung function. *J. Proteome Res.,* **2010**, *9*(6), 3083-3090.
[http://dx.doi.org/10.1021/pr1000048] [PMID: 20408573]

[26] Toromanović, J.; Kovac-Besović, E.; Šapčanin, A.; Tahirović, I.; Rimpapa, Z.; Kroyer, G.; Sofić, E. Urinary hippuric acid after ingestion of edible fruits. *Bosn. J. Basic Med. Sci.,* **2008**, *8*(1), 38-43.
[http://dx.doi.org/10.17305/bjbms.2008.2994] [PMID: 18318670]

[27] Folwarczna, J.; Janas, A.; Pytlik, M.; Cegieła, U.; Śliwiński, L.; Krivošíková, Z.; Štefíková, K.; Gajdoš, M. Effects of trigonelline, an alkaloid present in coffee, on diabetes-induced disorders in the rat skeletal system. *Nutrients,* **2016**, *8*(3), 133.
[http://dx.doi.org/10.3390/nu8030133] [PMID: 26950142]

[28] Lang, R.; Wahl, A.; Stark, T.; Hofmann, T. Urinary N-methylpyridinium and trigonelline as candidate dietary biomarkers of coffee consumption. *Mol. Nutr. Food Res.,* **2011**, *55*(11), 1613-1623.
[http://dx.doi.org/10.1002/mnfr.201000656] [PMID: 21618426]

[29] Chen, J.; Wang, W.; Lv, S.; Yin, P.; Zhao, X.; Lu, X.; Zhang, F.; Xu, G. Metabonomics study of liver cancer based on ultra performance liquid chromatography coupled to mass spectrometry with HILIC and RPLC separations. *Anal. Chim. Acta,* **2009**, *650*(1), 3-9.
[http://dx.doi.org/10.1016/j.aca.2009.03.039] [PMID: 19720165]

[30] Sykes, B.D. Urine stability for metabolomic studies: effects of preparation and storage. *Metabolomics,* **2007**, *3*(1), 19-27.
[http://dx.doi.org/10.1007/s11306-006-0042-2]

[31] Kochhar, S.; Jacobs, D.M.; Ramadan, Z.; Berruex, F.; Fuerholz, A.; Fay, L.B. Probing gender-specific metabolism differences in humans by nuclear magnetic resonance-based metabonomics. *Anal. Biochem.,* **2006**, *352*(2), 274-281.
[http://dx.doi.org/10.1016/j.ab.2006.02.033] [PMID: 16600169]

[32] Psihogios, NG; Gazi, IF; Elisaf, MS; Seferiadis, KI; Bairaktari, ET Gender-related and age-related urinalysis of healthy subjects by NMR-based metabonomics. *NMR in Biomedicine: An International Journal Devoted to the Development and Application of Magnetic Resonance In vivo.,* **2008**, *21*(3), 195-207.

[33] Lenz, E.M.; Bright, J.; Wilson, I.D.; Hughes, A.; Morrisson, J.; Lindberg, H.; Lockton, A. Metabonomics, dietary influences and cultural differences: a ^{1}H NMR-based study of urine samples obtained from healthy British and Swedish subjects. *J. Pharm. Biomed. Anal.,* **2004**, *36*(4), 841-849.
[http://dx.doi.org/10.1016/j.jpba.2004.08.002] [PMID: 15533678]

[34] Ewing, J.A. Detecting alcoholism. The CAGE questionnaire. *JAMA,* **1984**, *252*(14), 1905-1907.
[http://dx.doi.org/10.1001/jama.1984.03350140051025] [PMID: 6471323]

[35] Selzer, M.L. The Michigan alcoholism screening test: the quest for a new diagnostic instrument. *Am. J. Psychiatry,* **1971**, *127*(12), 1653-1658.
[http://dx.doi.org/10.1176/ajp.127.12.1653] [PMID: 5565851]

[36] Mostafa, H.; Amin, A.M.; Teh, C-H.; Murugaiyah, V.; Arif, N.H.; Ibrahim, B. Metabolic phenotyping of urine for discriminating alcohol-dependent from social drinkers and alcohol-naive subjects. *Drug Alcohol Depend.,* **2016**, *169*, 80-84.
[http://dx.doi.org/10.1016/j.drugalcdep.2016.10.016] [PMID: 27788404]

[37] Mostafa, H.; Amin, A.M.; Teh, C-H.; Murugaiyah, V.A.; Arif, N.H.; Ibrahim, B. Plasma metabolic biomarkers for discriminating individuals with alcohol use disorders from social drinkers and alcohol-naive subjects. *J. Subst. Abuse Treat.,* **2017**, *77*, 1-5.
[http://dx.doi.org/10.1016/j.jsat.2017.02.015] [PMID: 28476260]

[38] Zieve, D.; David, C. *Alcoholism and Alcohol Abuse*; NIH: Bethesda, **2011**.

[39] Luft, F.C. Lactic acidosis update for critical care clinicians. *J. Am. Soc. Nephrol.,* **2001**, *12* Suppl. 17, S15-S19.
[PMID: 11251027]

[40] Landaas, S.; Jakobs, C. The occurrence of 2-hydroxyisovaleric acid in patients with lactic acidosis and ketoacidosis. *Clin. Chim. Acta,* **1977**, *78*(3), 489-493.
[http://dx.doi.org/10.1016/0009-8981(77)90082-1] [PMID: 884872]

[41] Bowling, F.G.; Morgan, T.J. Krebs cycle anions in metabolic acidosis. *Crit. Care,* **2005**, *9*(5): E23.
[http://dx.doi.org/10.1186/cc3878] [PMID: 16277707]

[42] Calabrese, V.; Calvani, M.; Butterfield, D.A. Increased formation of short-chain organic acids after chronic ethanol administration and its interaction with the carnitine pool in rat. *Arch. Biochem. Biophys.,* **2004**, *431*(2), 271-278.
[http://dx.doi.org/10.1016/j.abb.2004.08.020] [PMID: 15488476]

[43] Lebouvier, T.; Chaumette, T.; Paillusson, S.; Duyckaerts, C.; Bruley des Varannes, S.; Neunlist, M.; Derkinderen, P. The second brain and Parkinson's disease. *Eur. J. Neurosci.,* **2009**, *30*(5), 735-741.
[http://dx.doi.org/10.1111/j.1460-9568.2009.06873.x] [PMID: 19712093]

[44] Magrinelli, F; Picelli, A; Tocco, P; Federico, A; Roncari, L; Smania, N. Pathophysiology of motor dysfunction in Parkinson's disease as the rationale for drug treatment and rehabilitation. *Parkinsons. Dis.,* **2016**, *2016*, 9832839.
[http://dx.doi.org/10.1155/2016/9832839]

[45] Shafique, H.; Blagrove, A.; Chung, A.; Logendrarajah, R. Causes of Parkinson's disease: literature review. *J Parkinsonism Restless Leg Syndr.*, **2011**, *1*(1), 5-7.
[http://dx.doi.org/10.7157/jprls.2011v1n1pp5-7]

[46] Cannon, J.R.; Greenamyre, J.T. Gene-environment interactions in Parkinson's disease: specific evidence in humans and mammalian models. *Neurobiol. Dis.*, **2013**, *57*, 38-46.
[http://dx.doi.org/10.1016/j.nbd.2012.06.025] [PMID: 22776331]

[47] Collier, T.J.; Kanaan, N.M.; Kordower, J.H. Ageing as a primary risk factor for Parkinson's disease: evidence from studies of non-human primates. *Nat. Rev. Neurosci.*, **2011**, *12*(6), 359-366.
[http://dx.doi.org/10.1038/nrn3039] [PMID: 21587290]

[48] Blesa, J.; Trigo-Damas, I.; Quiroga-Varela, A.; Jackson-Lewis, V.R. Oxidative stress and Parkinson's disease. *Front. Neuroanat.*, **2015**, *9*, 91.
[http://dx.doi.org/10.3389/fnana.2015.00091] [PMID: 26217195]

[49] Caudle, W.M.; Zhang, J. Glutamate, excitotoxicity, and programmed cell death in Parkinson disease. *Exp. Neurol.*, **2009**, *220*(2), 230-233.
[http://dx.doi.org/10.1016/j.expneurol.2009.09.027] [PMID: 19815009]

[50] Moon, H.E.; Paek, S.H. Mitochondrial dysfunction in Parkinson's disease. *Exp. Neurobiol.*, **2015**, *24*(2), 103-116.
[http://dx.doi.org/10.5607/en.2015.24.2.103] [PMID: 26113789]

[51] Rizek, P.; Kumar, N.; Jog, M.S. An update on the diagnosis and treatment of Parkinson disease. *CMAJ*, **2016**, *188*(16), 1157-1165.
[http://dx.doi.org/10.1503/cmaj.151179] [PMID: 27221269]

[52] Jankovic, J. Parkinson's disease: clinical features and diagnosis. *J. Neurol. Neurosurg. Psychiatry*, **2008**, *79*(4), 368-376.
[http://dx.doi.org/10.1136/jnnp.2007.131045] [PMID: 18344392]

[53] Ahmed, S.S.; Santosh, W.; Kumar, S.; Christlet, H.T.T. Metabolic profiling of Parkinson's disease: evidence of biomarker from gene expression analysis and rapid neural network detection. *J. Biomed. Sci.*, **2009**, *16*(1), 63.
[http://dx.doi.org/10.1186/1423-0127-16-63] [PMID: 19594911]

[54] Martin, E.; Rosenthal, R.E.; Fiskum, G. Pyruvate dehydrogenase complex: metabolic link to ischemic brain injury and target of oxidative stress. *J. Neurosci. Res.*, **2005**, *79*(1-2), 240-247.
[http://dx.doi.org/10.1002/jnr.20293] [PMID: 15562436]

[55] Sorbi, S.; Bird, E.D.; Blass, J.P. Decreased pyruvate dehydrogenase complex activity in Huntington and Alzheimer brain. *Ann. Neurol.*, **1983**, *13*(1), 72-78.
[http://dx.doi.org/10.1002/ana.410130116] [PMID: 6219611]

[56] Zhou, A.; Ni, J.; Xu, Z.; Wang, Y.; Lu, S.; Sha, W.; Karakousis, P.C.; Yao, Y.F. Application of (1)h NMR spectroscopy-based metabolomics to sera of tuberculosis patients. *J. Proteome Res.*, **2013**, *12*(10), 4642-4649.
[http://dx.doi.org/10.1021/pr4007359] [PMID: 23980697]

[57] Vignoli, A.; Tenori, L.; Giusti, B.; Takis, P.G.; Valente, S.; Carrabba, N.; Balzi, D.; Barchielli, A.; Marchionni, N.; Gensini, G.F.; Marcucci, R.; Luchinat, C.; Gori, A.M. NMR-based metabolomics identifies patients at high risk of death within two years after acute myocardial infarction in the AMI-Florence II cohort. *BMC Med.*, **2019**, *17*(1), 3.
[http://dx.doi.org/10.1186/s12916-018-1240-2] [PMID: 30616610]

CHAPTER 5

Applications of NMR Spectroscopy in Cancer Diagnosis

Asmaa A. Kamel[1] and **Fotouh R. Mansour**[2,3,*]

[1] Biochemistry Department, Faculty of Pharmacy, Tanta University, Tanta 31111, Egypt

[2] Department of Pharmaceutical Analytical Chemistry, Faculty of Pharmacy, Tanta University, Tanta 31111, Egypt

[3] Pharmaceutical Services Center, Faculty of Pharmacy, Tanta University, Tanta 31111, Egypt

Abstract: Cancer is a category of diseases characterized by uncontrolled cell growth and high potential to disseminate to other parts of the body. Cancer diagnosis is challenging due to the high structure similarity between normal and cancerous cells and the aggressive diagnostic procedures. Early diagnosis of cancer is crucial to increase the remission probability and avoid complications. A number of techniques have been involved in cancer diagnosis including biopsy, laboratory tests, computerized tomography (CT) scan, Ultrasonography, X-ray imaging, and nuclear magnetic resonance (NMR) spectroscopy. NMR has been applied both *in vivo* (known as magnetic resonance imaging) and *in vitro* to aid in cancer diagnosis. This chapter discusses the application of *in vitro* NMR in diagnosis and prognosis of different types of cancer with emphasis on the metabolic alterations at early stages of malignancy. The signature metabolites of brain, breast, epithelial ovarian, prostate, lung, colorectal, bladder, and oral cancers have been presented. A perspective overview of the role of NMR spectroscopy in cancer diagnosis has also been presented. This chapter shed the light on the important role of NMR spectroscopy in cancer diagnosis and treatment follow up. The applications introduced are not meant to provide a complete list of existing studies, but to present a wide overview of the current progress in this field. The chapter will cover the following topics:

Keywords: Applications, Bladder cancer, Brain cancer, Breast cancer, Cancer diagnosis, Colorectal cancer, Epithelial ovarian cancer, Lung cancer, Nuclear magnetic resonance (NMR) spectroscopy, Oral cancer, Perspective, Prostate cancer, Technical aspects.

---

* **Corresponding author Fotouh R. Mansour:** Department of Pharmaceutical Analytical Chemistry, Faculty of Pharmacy, Tanta University, Tanta 31111, Egypt; Tel: +20-10-6669-8099; Fax: +20-40-333-5466;
E-mail: fotouhrashed@gmail.com

Atta-ur-Rahman and M. Iqbal Choudhary (Eds.)
All rights reserved-© 2020 Bentham Science Publishers

INTRODUCTION

Nuclear magnetic resonance (NMR) spectroscopy is a powerful analytical technique for both identification and quantification of analytes in solutions as well as in solid states [1]. NMR phenomenon was discovered in 1940s [2] and since then, there has been a rapid progress with regard to both method development and applications, expanding from physics to chemistry, biochemistry, pharmacy, physiology, food science, biology, and medicine [3].

In the field of medical diagnosis, NMR spectroscopy provides a non-invasive metabolic window on the biochemical processes within the body [4]. Its use is no longer restricted to research to investigate pathophysiological processes, but extends to drug assessment, personalized medicine as well as biochemical characterization and diagnosis of diseases [5]. The use of NMR-based metabolomics to aid in human disease diagnosis would give a more complete picture as it reflects the integrated functions of organs [6]. Furthermore, metabolic changes can be detected in biological fluids using NMR spectroscopy before the clinical symptoms develop, generating useful fingerprints for early diagnosis of diseases [7].

NMR spectroscopy would also help in the challenge of cancer diagnosis, especially in brain tumor, by providing another non-invasive approach besides clinical history and radiological examination [8]. The additional metabolic information provided by NMR spectroscopy can help making clinical decisions about cancer patient management without surgical diagnostic procedure [9]. NMR spectroscopy also has a great impact on metabolite-based discovery of diagnostic and prognostic biomarkers of several human diseases [10]. More sensitive boimarkers are urgently needed because traditional biomarkers of diseases are not sensitive enough and only increase after the presence of substantial diseases [6].

The use of NMR spectroscopy for medical diagnosis can be conducted both *in vitro* and *in vivo*. The biomedical applications of *in vitro* NMR include the analysis of body fluids (such as plasma or urine), extracts of tissue or small biopsy-sized specimens of intact tissues [11]. On the other hand, *in vivo* NMR spectroscopy, commonly known as magnetic resonance spectroscopy (MRS), can be done on the whole-body using a clinical magnetic resonance imaging (MRI) scanner, as an adjunct to standard examination, to obtain metabolic and functional information complementary to anatomical changes [12].

This chapter focuses on the recent applications of NMR spectroscopy in medical diagnosis and how it could offer the potential for a holistic approach to clinical medicine *via* improving disease diagnosis, biomarkers discovery as well as understanding disease mechanisms. The selected applications provide a wide overview of the current progress in this field, and the future trends.

OVERVIEW OF NMR SPECTROSCOPY

Since NMR was first described in 1938 by Isidor Rabi, the applications and the number of publications are steadily growing [13]. Fig. (**1A**) shows the total number of publications (Journal articles, book chapters, patents, conference abstracts,) with the key word "NMR" using the Semantic Scholar search engine in the last two decades. NMR is one of the most widely used techniques in metabolomic studies (Fig. **1B**). The principle of NMR spectroscopy has been discussed in a number of text books [14, 15]. In this section, the types of NMR spectroscopy used in cancer diagnosis and the pros and cons of the technique will be discussed.

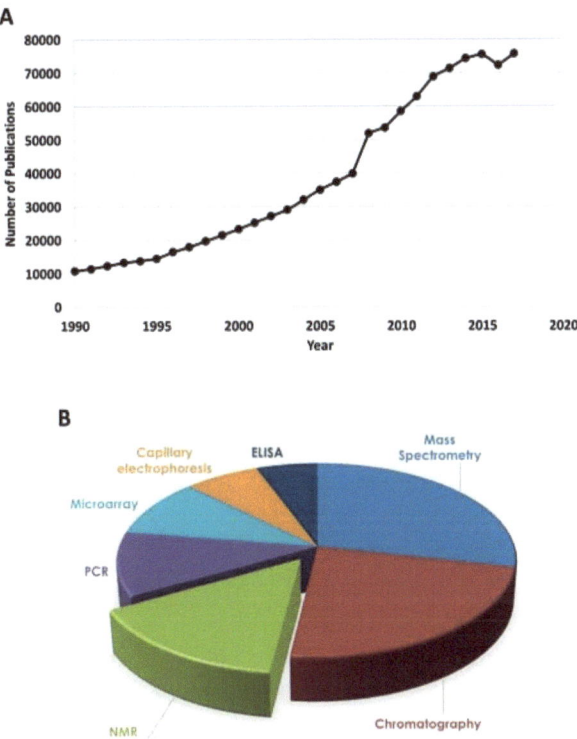

Fig. (1). A) The number of publications with keyword "NMR" in the last 20 years. **B)** A pie chart for the different techniques used in metabolomics.

Types of NMR Spectroscopy Used in Cancer Diagnosis

NMR can be classified according to the number of atoms ^1H, ^{13}C, ^{15}N, ^{19}F, ^{31}P, *etc*, where ^{1}H NMR is the most common type due to the high sensitivity compared with ^{13}C NMR and the wide applicability, because any organic compound has at least one proton while ^{19}F, ^{31}P are rarely found in metabolites. However, the small spectral width in ^{1}H NMR (0-10 ppm) makes the spectra overcrowded and usually overlapped, which may affect the reliability of the results. For this reason, 2D NMR spectroscopy are employed to simplify the spectrum and avoid signal overlapping. 2D NMR can correlate atoms of the same type (homonuclear) as well as of different types (Heteronuclear). Different variants of homonuclear 2D NMR have been utilized in metabolomics, including correlation spectroscopy (COSY), total correlation spectroscopy (TOCSY), diffusion ordered spectroscopy (DOSY) and 2D J-resolved NMR spectroscopy. The differences between these modes have already been discussed [16].

NMR can be classified according to the nature of the application into *in vivo* and *in vitro* NMR. Both *in vivo* and *in vitro* NMR apply a strong magnetic field and measure radio-waves to gain information. *In vivo,* NMR is commonly known as magnetic resonance imaging (MRI) and it is widely applied in clinical diagnosis to get images of the body organs and pathological processes. On the other hand, *in vitro* NMR are more common in metabolomics to study the metabolic changes in different diseases. In this chapter, only the application of *in vitro* NMR in cancer diagnosis will be discussed. Information about MRI and its applications in cancer can be found in other reviews [17 - 19].

Advantages and Disadvantages of NMR Spectroscopy

The number of Nobel prizes awarded for NMR specialists and the increasing number of applications every year are due to the several advantages of the technique [20]. NMR is applicable for a wide variety of analytes (hydrophilic, hydrophobic, organic, inorganic, solid, liquid or gas). NMR requires small sample sizes with minimal or no preparation. The results of NMR analysis are suitable for quantitative analysis, and the established library of NMR spectra is suitable for qualitative analysis. In addition, NMR could be the most powerful technique in structure elucidation due to the rich information gained and its non-destructive/non-invasive nature. The data obtained from NMR are highly reproducible and a prior sample separation is not essential [21].

However, NMR has some limitations, such as the low signal to noise ratio due to the high background response, which compromises the technique sensitivity. The spectra of biological samples are complicated and difficult to interpret, unless a mathematical treatment was performed. Finally, the high cost of the instrument which increases the cost of the sample analysis [22]. Yet, the ease of application and the wide scope of NMR explains why it is an attractive alternative to the more sensitive mass spectrometric (MS) technique. Table **1** shows a comparison of the strengths and weaknesses of NMR and MS. Best results are obtained when NMR results are coupled with MS data to get a comprehensive overview of the metabolic changes.

Table 1. comparison between strengths and weaknesses of NMR and MS (With permission from [23]).

	NMR	Mass spectrometry
Sensitivity	Low but can be improved with higher field strength, cryo and microprobes and dynamic nuclear polarization	High, but can suffer from ion suppression in complex and salty mixtures
Sample measurement	The entire sample analyzed in one measurement	Usually need different chromatography techniques for different classes of metabolites
Sample recovery	Nondestructive; sample can be recovered and stored for a long time, several analyses can be carried out on the same sample	Destructive technique but need a small amount of sample
Reproducibility	Very high	Moderate
Sample preparation	Need minimal sample preparation	More demanding; needs different LC columns and optimization of ionization conditions
Experimental time	5 min for 1D proton NMR	Less than 3 min for direct infusion but more than 10 min for simplest analysis by GC MS or LC MS
Tissue samples	Yes, using HRMAS NMR tissue samples analyzed directly	No, requires tissue extraction MS can be used to identify metabolites in tissues using MALDI-MS
Number of detectable metabolites in urine sample	40–100 depending on spectral resolution	Could be 500+
Target analysis	Inferior for targeted analysis	Superior for targeted analysis
In-vivo studies	Yes—widely used for 1H magnetic resonance spectroscopy (and to a lesser degree 31P and 13C)	No—although suggestion that DESI may be a useful way to sample tissues minimally invasively during surgery
Molecular dynamic, molecular diffusion	NMR can be used to probe the molecular diffusion and dynamics	No

TECHNICAL ASPECTS OF *IN VITRO* NMR APPLICATIONS

Human body is not a test tube; strict control over experimental conditions is not always possible, when collecting and handling biological samples for NMR analysis. Factors such as stress, exercise, diet, drugs or coexisting diseases can significantly affect the metabolic profiles of patients. For this reason, biological samples are better collected in the early morning after 12 to 16 hours of fasting, to avoid any effect of food, exercise or drugs [24]. Special attention should be paid to other health conditions of patients to avoid misleading conclusions or false positive results. Blood samples should be collected in tubes containing lithium heparin, as an anticoagulant [25]. EDTA and sodium citrate should be avoided due to possible overlap with NMR signals [25]. Plasma samples should be separated by centrifuging at 3000× g and 4°C for 7-15 minutes and collecting the supernatant liquid. Any delay between sample collection and certification is not recommended, to avoid over accumulation of lactate [26]. Short-term storage of plasma samples is possible at -20°C while long-term storage is recommended at -80°C. Storage at 4°C is not sufficient due to probable microbial growth and acetate formation [27]. Frozen samples should be gradually thawed to avoid degradation. Deuterated solvents should be used, otherwise strong background signals will be obtained due to solvent effects. Signals can be affected by pH changes, thus high concentrations of phosphate buffer (0.3-1 M, pH 7.4) are highly recommended. Strict standards operating procedures [27] should be followed to improve accuracy and precision of results (Fig. **2**).

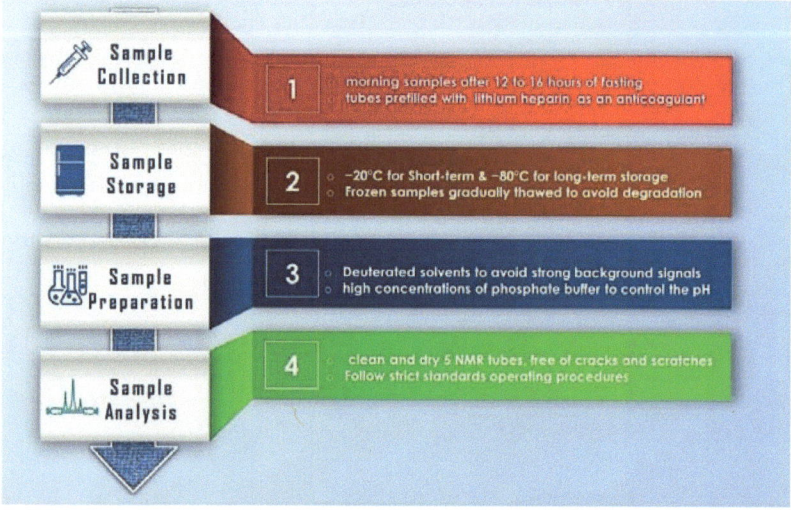

Fig. (2). Technical aspects of *in vitro* NMR spectroscopy during sample collection, storage, preparation and analysis.

APPLICATIONS

Brain Tumor

Survival rates vary considerably among brain tumor patients depending on the tumor origin (primary or secondary), type (glioma, meningioma, *etc.*), and grade (low or high). Accurate diagnosis of brain tumors improves the treatment decision and enhances the clinical outcome [28]. NMR spectroscopy can be applied to explore signature metabolites of brain tumors in blood, plasma and brain biopsies, revealing the status of tumor micro-heterogeneity and metabolic alterations even at an early stage of malignant transformation before they are morphologically detectable [29].

The metabolic profiles of HR-MAS ^{1}H NMR spectra, obtained from biopsies of human brain tumors have been used to generate predictive pattern recognition models to discriminate different tumor types. This approach allowed not only classification of tumors into meningeal and non-meningeal, but also glioma grading with an accuracy of 62% [23]. HR-MAS NMR spectra of different brain metastases compared by multivariate analysis could give additional information, along with the histopathological examination, about the biochemical composition of the tissue, which could improve the possibility of differentiating metastases according to origin, predict patient survival, and guide treatment decisions related to terminating further therapies. This finding is of high clinical interest, as the patient may suffer from brain metastases without a known history of primary tumor [30].

Each type of brain tumors has some selected biomarkers and metabolites that can be used for differential diagnosis. Major signals detected in ^{1}H NMR spectra of cerebral gliomas are those of choline-containing-compounds "tCho" (encompassing at least glycerophosphocholine (GPC), phosphocholine (PC), and choline (Cho)), total creatine "tCr" (Cr and phosphocreatine), lipids and other macromolecules, lactate, hydroxyglutarate, hypo-taurine, N-acetylaspartate (NAA) and N-acetylaspartyl glutamate (tNAA) [31]. Owing to the higher magnetic field strength of *in vitro* NMR systems, the sensitivity and spectral resolution are greatly improved, which increase the number of detectable metabolites including less concentrated ones such as alanine, glutamine or myo-inositol [32].

Besides differential diagnosis of the brain tumor type, tumor grading can be performed by NMR spectroscopy. When comparing metabolic NMR spectra of brain tumors, high-grade gliomas tend to have lower intensities of tNAA, proline, glutamate, glutamine, gamma amino butyric acid, and NAA and higher intensities of tCho, alanine, and valine than those of low-grade gliomas [33]. The concentrations of tNAA and tCr decrease in highly malignant gliomas to some extent, while being low or not detected in meningiomas. Higher levels of glutathione, glutamine, glutamate, and PC are observed in the atypical meningioma compared with the benign type [34].

Data interpretation of the literature is not straightforward as the diagnosis is not only based on absolute concentrations of metabolites, but also depends on metabolite ratios which are assumed to be related to the degree of malignancy. Multiple ratios have been used such as tCho/tCr, tNAA/tCr, tCho/tNAA, and inositol/tCr which have been predominantly used to grade brain tumors [31]. Major signature metabolites detected in brain tumors by NMR spectroscopy are shown in Table **2**.

Table 2. Major signature metabolites detected in brain tumors by NMR spectroscopy.

Cancer Type	Tumor Sample Source	Signature Metabolite/Identified Biomarker		Ref
		Increased	Decreased	
Brain	Plasma of gliomas patients vs healthy volunteers	LDL, unsaturated lipid, and pyruvate	Ile, Leu, Val, Lac, Ala, glycoprotein, Glu, citrate, Cr, MI, Cho, Tyr, Phe, 1-methylhistidine, α-glucose, and β-glucose	[35]
	biopsies of HGOs vs LGOs	Ala, and Val	Pro, Gln, Glu, GABA, and NAA	[33]
	biopsies of GBM vs grade II astrocytomas	PC, Gly, Ala, Tau, and lipids	Cr, MI, GPC, Cho, and Lac	[1, 10, 11]
	biopsies of GBM vs meningioma	Asp, Cr, GPC, His, MI, NAA, and SI	Ala, Glu, GSH, Ile, Tau, and Val	[36]
	biopsies of GBM vs metastasis	Cr, Gln, Gly, and H-tau	-----	[36]
	biopsies of grade II, III astrocytomas and GBM vs grade IV medulloblastomas	Lac, and Cr	MI, Tau, Gly, PC, GPC, and Asp	[28]
	biopsies of recurrent gliomas (IV vs II)	2-HG, H-tau, Ala, Cho, lipids, GSH, PC, and PC/GPC	MI	[12, 13]

Ala, alanine; Asp, aspartate; Cho, choline; Cr, creatine; GABA, γ-aminobutyrate; GBM, glioblastoma multiforme; Glu, glutamate; Gln, glutamine; GSH, glutathione; GPC, glycerophosphocholine; Gly, glycine; HGOs, high-grade oligodendrogliomas; 2-HG, 2-hydroxyglutarate; His, histidine; H-tau, hypotaurine; Ile, isoleucine; Lac, lactate; LDL, low-density lipoproteins; Leu, leucine; LGOs, low-grade oligodendrogliomas; MI, myo-inositol; NAA, N-acetyl- aspartate; PC, phosphocholine; Phe, phenylalanine; Pro, proline; SI, scyllo-inositol; Tau, taurine; Tyr, tyrosine; Val, valine.

Breast Cancer

Breast cancer (BC) is considered one of the worst killers of all cancers among women globally, accounting for 23% of all cancer cases and 14% of all cancer-related deaths, occurring one in every eight women [40]. Breast cancers are heterogeneous, and patients with apparently similar symptoms can have substantially different survival rates, responses to treatment, and treatment-related toxicity [41]. It is thus crucial to identify new approaches for BC prediction and early diagnosis, achieving a higher survival rate in more than three-quarters of BC patients [23].

NMR-based metabolomics has been proposed for early detection of BC. For example, HR-MAS NMR metabolomic analysis of biopsy samples has discriminated between BC patients and healthy controls with 69% sensitivity and 94% specificity, where higher levels of taurine and choline-containing metabolites were observed in cancer tissues compared with normal BC tissues [42].

Estrogen receptor (ER) and progesterone receptor (PgR) status is one of the most important prognostic factors in BC [34]. Classification of breast hormonal status based on HR-MAS NMR spectra of biopsies have been established with a prediction accuracy of 88%. ER-positive tumors contain more glycine, GPC, Cho, and alanine, while having less ascorbate, creatine, taurine, and PC when compared with ER-negative tumors. PgR-negative samples have more ascorbate, lactate, glycine, GPC, PC, choline, creatine, and alanine than PgR-positive tumors. These results suggest that the concentrations of altered metabolites combined with the ER and PgR levels could be potentially used as markers of hormonal receptor status to assess the usefulness of hormonal therapy of BC [43].

These metabolic alterations quantified by HR-MAS NMR were correlated with the histological type and size of the tumors and the patient's lymph node status. The concentrations of choline and glycine were found to be much higher in samples with tumors larger than 2 cm compared with samples with smaller tumors, suggesting that HR MAS NMR could help determination of BC stages [23]. Early detection of lymph node metastasis using both 1D and 2D proton NMR methods is also important for identification of high-risk patients and for establishment of personalized treatment [34]. Reduced glycine and increased taurine, GPC, PC and lactate were suggested to be major metabolic differences in lymph node–positive patients relative to lymph node–negative patients [44].

NMR metabolomic analysis of urine and blood samples served as a gateway for early diagnosis of BC, while analysis of breast tissue biopsy samples can be a useful tool as a form of secondary confirmation [45]. In a study by Slupsky *et al.* [46], NMR urinary metabolic profiling revealed that numerous metabolites decreased in concentration among BC patients when compared with the healthy group. The formate was ranked first among 67 metabolites by a significant decrease of 43% in BC patients. In comparison with mammography, which may result in multiple false positives and false negatives, NMR urinary metabolic profiling is faster, less costly, non-invasive, and more efficient (100% sensitivity and 93% specificity) for early BC screening [7].

In a recent study by using NMR-based serum metabolomics analysis, Singh *et al.* [47] have demonstrated that patients with over-expressed inositol 1, 4, 5 trisphosphate receptor (IP3R) were found to have higher serum levels of lactate, lysine, and alanine, while having lower levels of pyruvate and glucose compared to healthy individuals, which indicated that all of these serum metabolites could be identified as biomarkers for BC as shown in Fig. (**3**). Jobard *et al.* [48] have also used the NMR-based metabolomics approach to analyze serum of patients with metastatic BC and early BC, and it was found that the levels of acetoacetate, glycerol, pyruvate, N-acetyl glycoproteins, mannose, glutamate, and phenylalanine increased, while the concentration of histidine decreased in metastatic BC relative to early BC, suggesting that NMR metabolic signature of serum may have an important role in the diagnosis of BC metastasis, with a sensitivity of 89.8% and a specificity of 79.3%.

NMR serum metabolomic signature was also applied for the prediction of the relapse of early BC. A random forest (RF) classification model based on preoperative NMR-based serum metabolomic profiles was proposed by Hart *et al.* [49] as a new tool to predict the recurrence of ER-positive early BC with an accuracy of 71.3%. In another study by Tenori *et al.* [50], it was suggested that serum metabolomic analysis using ^{1}H-NMR spectroscopy may play a role in predicting the treatment efficacy and the overall survival of patients with HER2-positive BC treated with paclitaxel plus lapatinib.

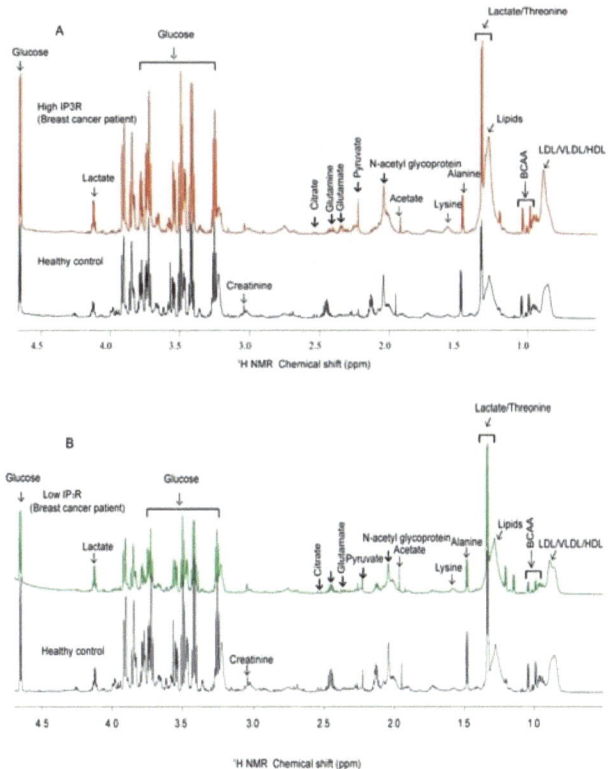

Fig. (3). Stack plot of representative ^{1}H NMR CPMG spectra of serum from A) healthy control and patient with high tissue expression of IP3R, B) healthy control and patient with low tissue expression of IP3R. All the spectra were plotted at same vertical scale for quantitative comparison (Reprinted with permission from [47]).

Ovarian and Endometrial Cancer

Epithelial ovarian cancer (EOC), the most common form of malignant ovarian tumor, is the fifth leading cause of cancer-related death among women in the western world [51]. The high mortality rate of EOC occurs primarily because the early-stage disease (I/II) is usually asymptomatic so more than 75% of the affected women are diagnosed at advanced stages (III/IV) [52]. Early detection of EOC while still located in the ovaries could yield a survival rate of 93% as opposed to only 15-20% if diagnosed at late stages [53]. The cancer antigen 125 and transvaginal ultrasound are two major powerful approaches used to diagnose EOC currently, however, they have low positive predictive values and are responsible for predicting less than 10% of EOC [54]. Therefore, besides genomic and proteomic analyses, there has been a great interest in the applications of NMR-based metabolomics to discover novel validated and more discriminating biomarkers to allow early and effective screening of EOC [23].

In a pioneering study using proton NMR analysis of ovarian cyst fluid samples from 40 patients with ovarian tumor, Boss *et al.* [55] have detected significant differences in the concentration of 36 metabolites between malignant and benign ovarian cysts. Since then, a number of studies have attempted to use NMR-based metabolomics for the analysis of a readily accessible body fluid, such as urine or serum as an early diagnosis approach. Odunsi *et al.* [56] have found that ^{1}H-NMR spectral differences could discriminate sera of women with EOC from those of healthy postmenopausal controls with 100% sensitivity and specificity as shown in Fig. (**4**). The same group has also shown in another study that proton NMR of a patient's serum allowed the diagnosis of early-stage I/II EOC which could significantly affect the clinical outcome of EOC patients [57]. Slupsky *et al.* [46] have also shown that among 67 metabolites detected *via* NMR-based urinary metabolic profiling, the singlet of methanol found at 3.35 ppm was ranked as the most important metabolite for distinguishing the ovarian cancer patients with 65% decrease in concentration relative to normal subjects.

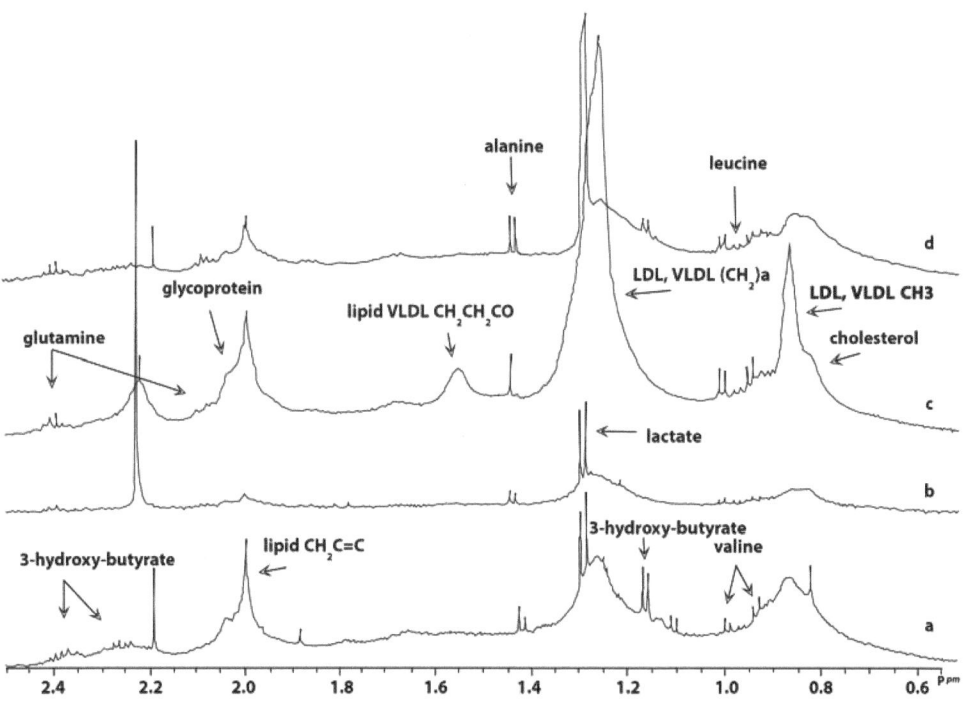

Fig. (4). Comparison of patients with epithelial ovarian cancer (EOC) with healthy subjects. The 600 MHz ^{1}H-NMR spectra of serum samples from a postmenopausal stage I EOC patient (**a**), a premenopausal healthy subject (**b**) a postmenopausal healthy subject (**c**) and a patient with benign ovarian cyst (endometriosis) (**d**) (Reprinted with permission from [56]).

The ability to detect circulating tumor cells (CTCs) with high sensitivity could help early detection of cancer and monitoring of treatment response. Most current technologies use EpCAM expression as a CTC identifier, however, during epithelial-to-mesenchymal transition, CTCs may express little or no EpCAM [58]. A novel sensing technology termed micro-nuclear magnetic resonance (μNMR) has been established which exploits magnetic resonance to detect cells labeled with magnetic nanoparticles. By comparing the performance of the quad-μNMR system to EpCAM expression, Ghazani *et al.* [59] have reported that quad-μNMR system could detect higher counts of circulating tumor cells in patients with advanced ovarian cancer (stages III, IV) and in patients not pursuing active therapy. The use of quad-μNMR would thus be especially helpful for detecting CTCs regardless EpCAM expression levels.

Malignant ascites is a major cause of morbidity that occurs in 37% of ovarian cancer patients. High-resolution ^{1}H NMR metabolic markers of malignant ascites can be used to improve characterization and diagnosis of ovarian cancer. Distinct metabolite patterns were detected in ascitic fluid collected from ovarian tumor-bearing mice that were not reflected in the corresponding cell culture or conditioned medium [60].

Endometrial cancer (EC) is the most common gynecologic malignancy in the developed world and affects more than 319,000 women worldwide each year [61]. EC is characterized by significant disturbances in crucial cellular metabolic pathways. Bahado-Singh *et al.* [62] have detected extensive changes of 32 metabolites in the sera of EC patients compared to healthy controls by using NMR metabolomics analysis approach, and some of these metabolites could be used to accurately predict early-stage EC. The results suggested that 3-hydroxybutyrate, hexanoylcarnitine, or tetradecadienyl-l-carnitine could serve as biomarkers to predict EC.

Prostate Cancer

Prostate cancer (PCa) is one of the major threats to men's health worldwide. PCa is ranked as the second most frequent cancer and the fifth leading cause of cancer death in men [63]. The current diagnostic approaches for PCa include prostate-specific antigen (PSA) or digital rectal examination, confirmed by a prostate biopsy for determination of the prognosticator Gleason score. However, these methods lack both specificity and sensitivity [64]. Moreover, some prostate biopsies may miss a significant amount of small cancers due to prostate tumor heterogeneity, resulting in underdetection of higher-grade (clinically significant) PCa or overdetection of low-grade (clinically insignificant) cancers, highlighting the need for other biomarkers [65]. This overdetection can lead to overtreatment of patients by receiving unnecessarily aggressive interventions or undergoing repeated assessment over time, which overburdens patients and health care systems [66].

An alternative diagnostic modality involves the use of NMR spectroscopy to accurately differentiate between healthy individuals, PCa patients, and benign prostatic hyperplasic (BPH) patients [67]. In a study by Zaragoza *et al.* [68], 113 urine samples were tested for PCa using ^{1}H-NMR spectroscopy, where it was suggested that the presence of PCa cannot be detected by a single analyte, but it is rather associated with alterations in the concentration of multiple metabolites, including phosphocholine, myo-inositol, spermine, glutamine, citrate, alanine, lactate, OH-butyrate, valine, and leucine. Pérez-Rambla *et al.* [69] have documented that ^{1}H-NMR urine metabolomic profile of PCa patients is characterized by increased concentrations of branched-chain amino acids, glutamate and pseudouridine, as well as decreased concentrations of glycine, dimethylglycine, fumarate and 4-imidazole-acetate, compared with patients with BPH.

The advances in NMR spectroscopy have been applied for diagnosing and staging PCa with higher accuracy compared to the current clinical diagnostic methods. Using NMR-based metabolomics analysis of serum by Kumar *et al.* [70], it was found that increased levels of alanine, pyruvate, and sarcosine and decreased levels of glycine were able to differentiate 90.2% of PCa patients from healthy people with 84.4% sensitivity and 92.9% specificity. Moreover, high-grade PCa and low-grade PCa could also be discriminated by the combination of three biomarkers (alanine, pyruvate and glycine) with 92.5% sensitivity and 93.3% specificity. The same group, used a serum ^{1}H-NMR-based metabolomics approach, to monitor the alteration in the concentrations of alanine, sarcosine, creatinine, glycine and citrate, which could be used to differentiate between BPH and PCa, with an accuracy of 88.3% (compared to an accuracy of 75.2% by clinical laboratory method). Furthermore, glycine, sarcosine, alanine, creatine, xanthine and hypoxanthine were used to distinguish abnormal prostate (BPH and PCa) from healthy individuals, with high precision of 86.2% (compared to a precision of 68.1% by clinical laboratory method) [71].

The intact prostate tissue samples were also analyzed using HR MAS NMR spectroscopy to reveal the overall tumor pathologic status to precede histologically observable changes in cell morphology [72]. It was proposed that spermine can be used as an endogenous marker of prostate cancer growth, because HR-MAS shows a correlation between the spermine concentration and the %volume of normal prostatic epithelial cells, as measured by histopathology [34]. Beger [73] has also reported that metabolic ratios of (glycerophosphocholine and phosphocholine)/creatine, myo-inositol/scyllo-inositol, and choline/creatine were correlated with the number of tumor cells and tumor cell proliferation.

Prostate tumor comprises aggressive and indolent varieties. Accurate diagnosis is a real challenge to discriminate between these varieties, to determine the prognosis, and to reduce the risk of overtreatment [74]. It was reported by Giskeødegård et al. [75] that metabolic profiling of prostatectomy tissue can be used to identify specific metabolites as biomarkers for aggressiveness. After the analysis of 158 prostate tissue samples from 48 patients by HR-MAS, it was concluded that high- and low- grade cancer tissues could be distinguished by decreased concentrations of spermine and citrate, and an increase in (total choline+creatine+polyamines)/citrate (CCP/C) ratio. The metabolic profiles were significantly correlated to the Gleason score obtained from each tissue sample with a sensitivity of 86.9% and a specificity of 85.2%. In a retrospective study of Vandergrift et al. [76], the HRMAS-metabolic profile analysis of PCa showed that myo-inositol was elevated in patients with highly aggressive cancers. It could also confidently identify sub-groups of patients with less aggressive PCa, allowing the use of personalized medicine.

Prostatic and seminal fluids can also be employed to search for PCa biomarkers. The prostatic fluid is more suitable than the seminal fluid because the seminal fluid may contain contaminants from other organs, including testes and seminal vesicles [77]. Pure prostatic fluids can be obtained by prostate massage, while the seminal fluid can be collected by ejaculation [78]. Seminal plasma samples of 151 men have been investigated for PCa with ^{1}H-NMR spectroscopy by Roberts et al. [79], where it was found that metabolomic signature of seminal plasma may assist diagnosis and monitoring of either low- or intermediate- grade PCa, whereas it shows less clinical benefit for diagnosing high-risk patients. Serkova et al. [80] have used ^{1}H-NMR spectroscopy to assess potential metabolic markers of PCa in human expressed prostatic secretions obtained from 52 men with PCa and from 26 healthy controls. It was concluded that the absolute concentrations of citrate, myo-inositol, and spermine were highly predictive of PCa and inversely related to the risk of PCa as shown in Fig. (**5**). At 90% sensitivity, these metabolites had specificities of 74%, 51%, and 34%, respectively. Furthermore, the obtained metabolic profile was independent on age, increasing the possible utility of these markers, after elimination of age as an interfering variable.

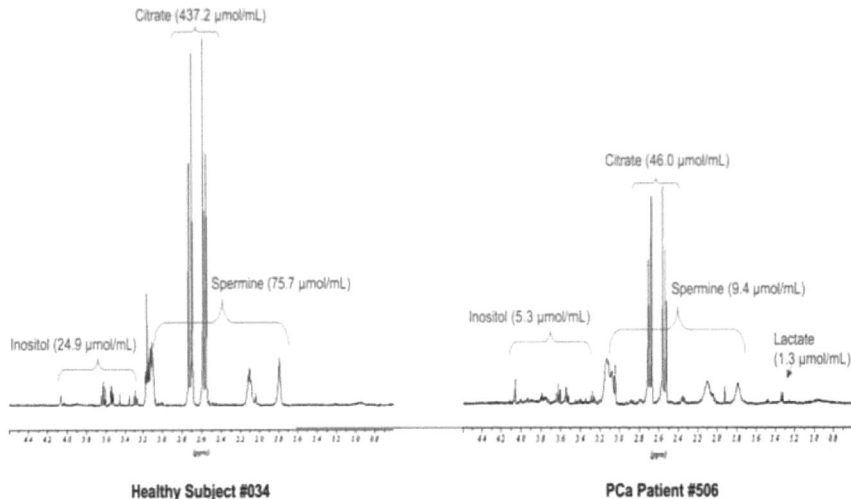

Fig. (5). Two representative high-resolution NMR spectra from human expressed prostatic secretions from a healthy control subject (left) and a PCa patient (right, 5 magnification) (Reprinted with permission from [80]).

Lung Cancer

Lung cancer (LC) is the leading cause of cancer incidence and mortality worldwide. LC accounted for about 2.1 million new cases and 1.8 million deaths in 2018, with less than 15% five-year survival rate [63]. Most LC patients (>80%) are diagnosed at late stages, when they are no longer amenable to surgery, leading to a poor prognosis associated with high mortality rates [81]. The early detection of LC before metastasizing to lymph nodes or distant sites would increase the five-year survival rates drastically up to 80% [23]. However, the majority of LC patients show no signs or symptoms at the initial stages and there are no established biomarkers for the early diagnosis [82]. Moreover, the currently available diagnostic approaches, such as computed tomography, MRI, and positron emission tomography are not suitable for general screening due to their high costs and the little information they provide about predictors of cancer progression [83]. Therefore, it is crucial to find complementary diagnostic approaches which allow an earlier detection of LC, to enhance patient management and prognosis [84].

NMR spectroscopy is a promising approach to find reliable biomarkers in different biological samples including plasma, urine, lung tissue, sputum and exhaled breath condensate (EBC) at earlier tumor stages, allowing it to be used as a screening modality to identify suspicious cases for subsequent and more specific radiological tests. Rocha *et al.* [82] have discriminated 85 patients with primary

LC from 78 healthy subjects via ^{1}H-NMR plasma-based metabolomics with sensitivity and specificity levels of about 90%. LC patients at initial disease stages showed significant higher levels of low-density lipoproteins (LDL), very low-density lipoproteins (VLDL), lactate and pyruvate and lower concentrations of HDL, glucose, citrate, formate, acetate, methanol, and several amino acids (alanine, glutamine, histidine, tyrosine, valine compared to the control group. Louis et al. [85] have also reported that ^{1}H-NMR metabolic profiles of plasma can be used to distinguish the LC patients from healthy individuals with a sensitivity of 71% and a specificity of 81%. The observed metabolic changes which can be detected at very early stages of LC involve higher levels of glucose and lower levels of lactate and phospholipids in LC patients compared to the control group.

Urine samples have been analyzed using NMR spectroscopy in order to identify possible biomarkers with a potential diagnostic value. In a study by Carrola et al. [84], 71 LC patients were distinguished from 54 healthy individuals using ^{1}H-NMR urinary-based metabolomics with 93% sensitivity, 94% specificity, and 93.5% classification rate. The resulted metabolic profile showed that the concentrations of creatinine, trimethylamine N-oxide/betaine, hippurate, citrate, α-ketoglutarate, glycine, hippurate, and trigonelline were decreased, whereas β-hydroxyisovalerate, α-hydroxyisobutyrate, N-acetylglutamine, and creatinine were increased in the LC patients compared to the control group. The authors proposed that the possible confounding factors, including gender, age, and smoking showed a minimal influence on this discrimination, but the contribution of other factors such as diet should be adequately evaluated in future studies.

HR-MAS NMR spectroscopy was used in combination with the multivariate data analysis to provide a more realistic insight into the metabolic profiles of lung tissues. Rocha et al. [86] have analyzed paired samples of tumor and noninvolved adjacent tissues from 12 lung tumors using ^{1}H-HR-MAS NMR . The results showed that lactate, PC, and GPC were elevated in tumors, while glucose, myo-inositol, inosine/adenosine, and acetate were reduced relative to the control tissues. Chen et al. [87] have also detected the metabolomic differences of 51 lung tissues at different sites from 17 patients with LC using the ^{1}H-HR-MAS NMR spectroscopy. Compared to the adjacent noninvolved tissues, the LC tissues showed significantly higher levels of aspartate, PC, GPC, and lactate but significantly lower levels of glucose and valine. In another study, Chen and his colleagues have utilized both ex vivo ^{1}H-HR-MAS NMR and in vitro ^{1}H-NMR spectroscopy techniques synchronously to analyze the metabolomic signature of 102 lung tissues from 34 patients with LC to identify potential diagnostic tissue biomarkers [81]. It was found that the concentrations of lipids and lactate were significantly increased, while the levels of myo-inositol and valine were declined in the cancer tissues compared with the adjacent non-involved tissues.

Novel metabolic biomarkers of LC were also identified in sputum and EBC samples using NMR spectroscopy. Ahmed *et al.* [88] have collected sputum and EBC samples from 20 patients (half of them with LC while the remaining patients have benign respiratory conditions). Sputum samples were further confirmed cytologically to differentiate between true sputum and saliva. The EBC samples of LC patients showed lower methanol levels and higher concentrations of propionate, ethanol, acetate, and acetone compared to the patients with benign conditions. The cytologically confirmed sputum and the combined sputum and saliva samples of LC patients were free of glucose and showed lower concentrations of *N*-acetyl sugars, glycoprotein, propionate, lysine, acetate, and formate compared to patients with benign conditions. It was supposed that the absence of glucose in sputum and lower concentrations of methanol in EBC of LC patients obtained by ^{1}H-NMR may serve as metabolic biomarkers of LC for early detection, monitoring treatment response, and detecting recurrence. The ^{1}H-NMR spectra of sputum as well as EBC from the LC patients and the control subjects are shown in Figs. (**6** and **7**), respectively.

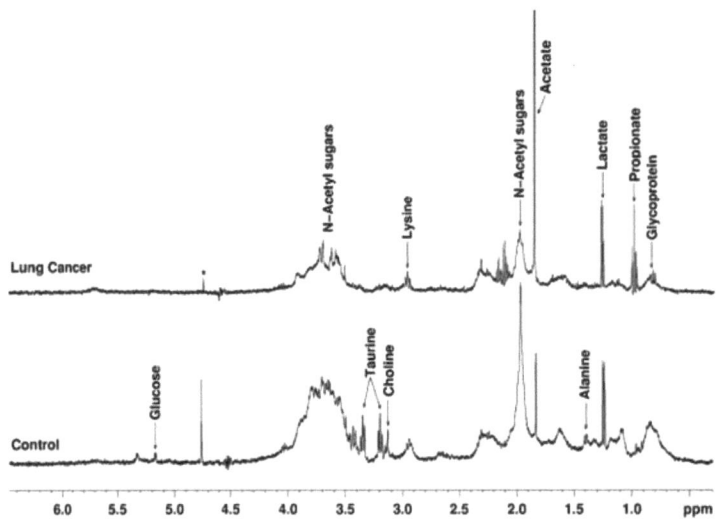

Fig. (6). ^{1}H-NMR spectra of sputum samples from a control subject and a lung cancer patient showing relative levels of metabolites including the absence of glucose in lung cancer patient (Reprinted with permission from [88]).

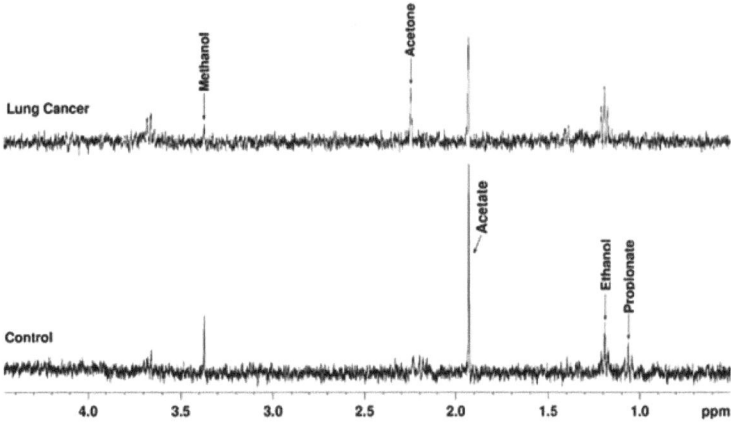

Fig. (7). ¹H-NMR spectra of exhaled breath condensate (EBC) samples from a control subject and a lung cancer patient showing relative levels of metabolites (Reprinted with permission from [88]).

Lung cancers can be classified into small cell lung carcinoma and non-small cell lung carcinoma (NSCLC) where NSCLC includes adenocarcinoma (AdC) and squamous cell carcinoma (SqCC). These categories are helpful for planning treatment and determining prognosis which differ significantly by stage, so it is a critical diagnostic challenge to discriminate between different LC subtybes [83]. In a study of Puchades-Carrasco *et al.* [89], the NMR blood-based metabolic profiles of NSCLC patients were distinguished from that of the healthy control group on the basis of significant changes in the concentration of 18 metabolites, including lipids, different amino acids, organic acids and alcohols. In addition to that, another 17 metabolites were found to be involved in metabolic changes in different stages of disease progression.

In another study of Rocha *et al.* [90], HR-MAS NMR spectroscopy was used to analyze metabolic signatures of matched tumor and adjacent control tissues from 56 patients undergoing surgical excision of primary LC. The spectra obtained in patients with AdC were characterized by increased levels of taurine and uridine nucleotides together with major alterations in phospholipid metabolism (increased PC, GPC, and phosphoethanolamine, as well as decreased acetate) and protein catabolism (increased peptide moieties). On the other hand, SqCC showed stronger glycolytic and glutaminolytic profiles (negatively correlated variations in glucose and lactate and positively correlated elevations in glutamate and alanine) together with increased concentrations of creatine and glutathione. The multivariate data analysis allowed differentiation between tumor and control tissues with high accuracy (97% classification rate), and also helped discrimination between AdC and SqCC profiles with a 94% classification rate, indicating the great potential for subtyping of LC.

The alterations of some important metabolites in LC tissues were found to be strongly correlated with the different stages of LC. After the analysis of lung tissues from 32 LC patients using ^{1}H-HR-MAS NMR, Chen *et al.* [91] have concluded that the increase of tumor staging was associated with elevations in the concentrations of PC, GPC, and lipids in tumor tissues. However, these metabolites were declined when the tumor progresses toward higher stages (III or IV). Aspartate was also found to be significantly increased in LC tissues with the increases of tumor staging, providing a good separation between the different classes. In another study, using ^{1}H-NMR spectroscopy of serum specimens obtained from 25 non-metastatic LC patients, Hao *et al.* [92] were able to distinguish LC at stages (1–2) from stage 3, based on 8 metabolites (2-hydroxybutyrate, 2-oxoisocaproate, acetate, carnitine, 3-hydroxyisovalerate, 2-hydroxyisovalerate, glycerol and glycine) with the help of scores they have established; patients at lower stages had a lower score in comparison to patients with higher staging as shown in Fig. (**8A**). They have also differentiated between AdC and SqCC using a score based on 19 abundant spectral differences as shown in Fig. (**8B**). The study results showed that patients with AdC had relatively lower levels of pyruvate, lactate, valine, proline, isoleucine, histidine, 2-aminobutyrate, leucine and alloisoleucine, and increased concentrations of 2-oxoisocaproate, 4-hydroxybutyrate, lysinearginine, dimethylamine, isobutyrate, 3-hydroxybutyrate, acetate, asparagine, phenylalanine compared to SqCC patients.

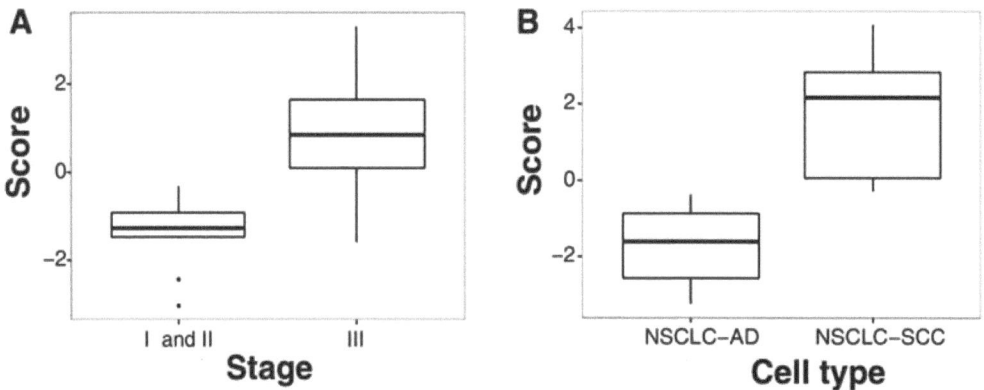

Fig. (8). NMR metabolite bioprofiling facilitates evaluation of pathological tumor characteristics. **A:** Box and whisker plot reflective of tumor staging: Scores from multivariate data analysis after baseline samples were stratified into stages 1 and 2, *versus* 3. **B:** Box and whisker plot based on cancer cell type with sample stratified as non-small cell lung cancer type squamous cell and adenocarcinoma (Reprinted with permission from [92]).

Chronic obstructive pulmonary disease (COPD) may increase the risk of developing LC, so it is crucial to distinguish between these two related pathological condition, especially considering the early stages of lung cancer. In a study by Deja et al. [93], serum samples were collected from 22 patients with COPD and 77 patients with NSCLC at different stages, then analyzed using NMR spectroscopy. The results showed that all LC patients had lower levels of acetate, citrate, and methanol together with elevated N-acetylated glycoproteins, leucine, lysine, mannose, choline, and lipid levels compared with the COPD group. The authors have also observed that the advanced NSCLC patients had higher concentrations of glycerol and *N*-acetylated glycoproteins and lower concentrations of isoleucine and acetoacetate compared to the early NSCLC patients, so these metabolites have the potential to stage NSCLC.

Colorectal Cancer

Colorectal cancer (CRC) is one of the most prevalent types of cancer worldwide, with more than 1,000,000 new cases and 500,000 deaths occurring annually [63]. The early diagnosis of localized pre-invasive CRC can markedly increase the five-year survival rates up to 95% [94]. The currently available screening and detection modalities for CRC involve colonoscopy, sigmoidoscopy, CT colonography, and histologic examination [95]. Although colonoscopy remains the gold standard to diagnose CRC, it is invasive, expensive, and uncomfortable [96]. CT colonography is also limited by its radiation hazard and the high cost [97]. The other non-invasive diagnostic tests include fecal occult blood test (FOBT), fecal immunochemical test (FIT), fecal-based DNA test, and blood-based DNA test (SEPT9 assay) [98]. FOBT and FIT are convenient methods for CRC screening, but they lack sensitivity and reliability. Fecal DNA and SEPT9 tests have relatively higher sensitivity, but their current cost is high for a screening assay [99]. These limitations highlight the need for an accurate, noninvasive, and inexpensive tool for early diagnosis of CRC [100].

NMR metabolic analysis of CRC patients' serum or urine may act as a preliminarily screening test before the invasive examination. Based on ^{1}H-NMR serum analysis of CRC patients, Zamani et al. [101] have identified 15 metabolites that changed markedly, such as pyridoxine, orotidine and taurocholic acid. These altered metabolites are mainly involved in bile acid biosynthesis, vitamin B6 metabolism, methane metabolism and glutathione metabolism which may explain the importance of lowering serum lithocholic acid/deoxycholic acid ratio and increasing vitamin B6 intake to help prevent colon cancer. Therefore, NMR spectroscopic analysis of serum would reveal the important CRC-induced metabolic aberrations that may further help in disease prevention. Wang et al. [102] have also used ^{1}H-NMR to profile urine metabolites from 55 CRC patients and 40 healthy controls. Unique metabolomic signatures have differentiated the urine samples of all CRC stages from those of healthy controls.

The local metabolic variations in the tissue biopsies of CRC patients can be identified using the HR-MAS NMR technique [34]. In the study of Mirnezami *et al.* [103], tissue samples obtained from both tumor center and margin were analyzed by HR-MAS NMR to develop novel metabolite-based model for diagnosis of CRC. A total of 171 spectra were obtained, including significantly increased levels of lactate, taurine, and isoglutamine as well as decreased levels of lipids/triglycerides in cancer tissues relative to healthy mucosa as shown in Fig. (**9**). The same group have also applied HR-MAS NMR spectroscopic profiling to differentiate between colonic and rectal cancers where colon cancer samples contained higher levels of acetate and arginine and lower levels of lactate relative to rectal cancer samples as shown in Fig. (**10**).

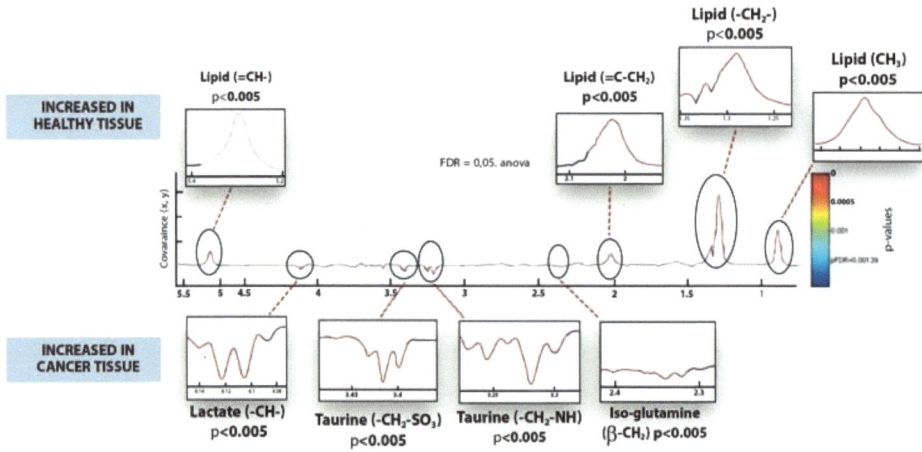

Fig. (9). Loadings plot allowing identification of the main metabolites responsible for discrimination between CRC and healthy tissues in generated OPLS-DA model (Reprinted with permission from [103]).

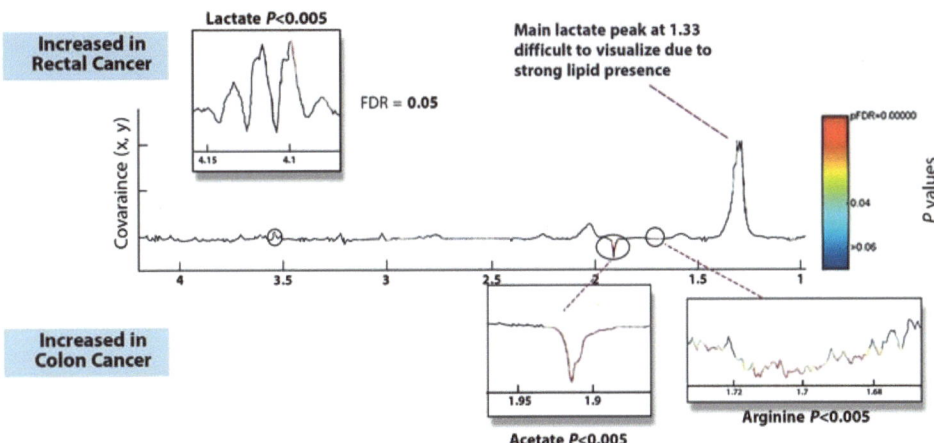

Fig. (10). HR-MAS NMR spectral metabolite pattern differences between rectal cancer and colon cancer (Reprinted with permission from [103]).

Detection of lymph node metastasis is a decisive prognostic factor for CRC that helps predict the overall survival (OS) and guide the therapeutic regime [42]. In a study of Bertini *et al.* [104], serum samples from 153 patients with metastatic CRC (mCRC) and 139 healthy subjects were analyzed using ^{1}H-NMR, where 96.7% of subjects were correctly classified with a cross-validated accuracy of 100%. The prognostic classification capabilities of ^{1}H-NMR profiling were then tested using a model built by selecting a subset of 20 patients with mCRC with maximally divergent OS: that is, the 10 patients with the shortest OS and the 10 patients with longest OS were studied. The set model was then used to determine the OS of 108 mCRC patients, where 85 patients were predicted to have long OS, while 23 patients were found to have short OS with an accuracy of 78.5%. In another study by Zhang *et al.* [105], ^{1}H-NMR was applied to determine the metabolic fingerprinting of tissue samples obtained from lymph node non-metastatic CRC patients (n=73), lymph node metastatic CRC patients (n=52) and normal controls (n=41). The results showed that 42 specific metabolites allowed not only differentiation of CRC patients from normal controls, but also provided excellent classification of CRC patients according to lymph node metastasis as presented in Fig. (**11**).

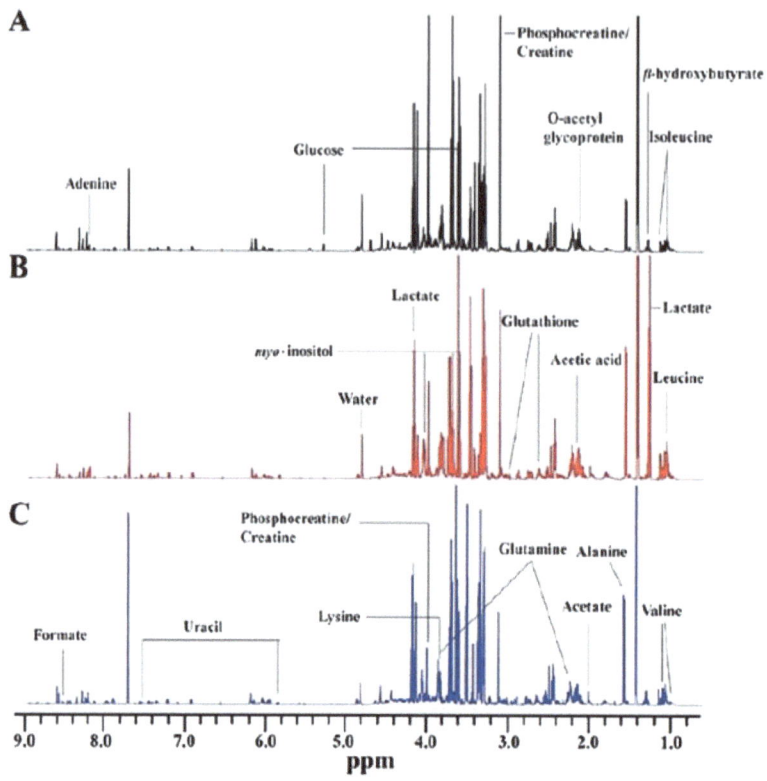

Fig. (11). Representative 600 MHz ^{1}H-NMR spectra of the tissue samples. **(A)** Normal control sample, **(B)** non-metastatic CRC sample, **(C)** metastatic CRC sample (Reprinted with permission from [105]).

Determining the correct clinical stage of CRC is of the utmost importance because treatment options vary greatly depending on the aggressiveness of disease at diagnosis [106]. NMR spectroscopy was found to be able to accurately detect the clinicopathologic stage of CRC using tumor tissue, mucosa adjacent to tumor or serum samples. In a study by Jiménez et al. [107], 83 intact tumor samples and 87 specimens of adjacent mucosa (10 cm from the tumor margin) obtained from 26 patients undergoing surgical resection for CRC were analyzed using HR-MAS NMR metabolic profiling. It was found that tumor adjacent mucosa showed unique metabolic changes that distinguish between tumors of different T- and N-stages according to TNM classification with higher predictive capability than tumor tissue itself as shown in Fig. (**12**). In addition, the adjacent off-tumor tissue samples could accurately predict the 5-year survival, offering a highly novel way of tumor classification and prognostication in CRC. NMR-based analysis of serum samples can also be used instead of the invasive biopsy to determine the stage of CRC [72]. In the study of Vahabi et al. [108], NMR-based metabolomics analysis of 16 patients' serum samples (8 at stage 0–I and 8 at stage II–IV) revealed that patients at stage II–IV had higher levels of glycine, cholesterol, taurocholic acid, cholesteryl ester and deoxyinosine and lower concentration of pyridoxine compared to patients at stage 0 and I.

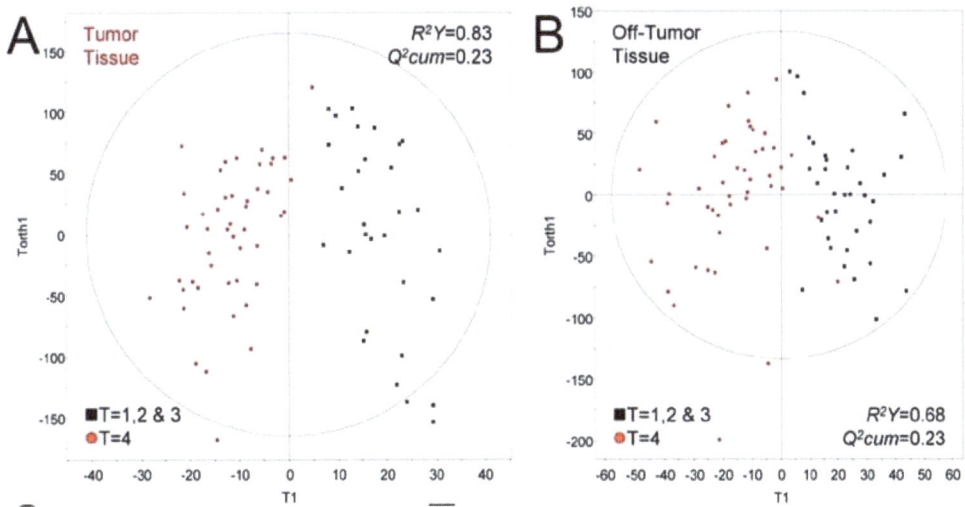

Fig. (12). CRC T-stage classification using HR-MAS metabolic profiles of tumor tissue and adjacent off-tumor tissue. Scatter plots for the OPLS models built using **(A)** tumor and **(B)** off-tumor tissue: the model shows the predictability for tumors classified as T = 4 (red ●) compared to those classified otherwise, T = 1–3 (■) (Reprinted with permission from [107]).

Although serum and urine samples have been widely analyzed by NMR-based metabolomics for CRC detection, the analysis of human feces seems to be more effective because feces are anatomically attached to the colorectal epithelium and carries metabolites derived from gut microbial-host co-metabolism [109]. In addition, large amount of exfoliated cells from the colonic mucosa and colorectal tumor can be shed into the feces, providing a rich source for detecting tissue-specific metabolic biomarkers of CRC [100]. Amiot *et al.* [110] have investigated the fecal metabolic profiling of patients with advanced colorectal neoplasia and controls using ^{1}H-NMR. Fecal concentrations of four short-chain fatty acids (valerate, acetate, propionate, and butyrate) were elevated, while β-glucose, glutamine, and glutamate levels were decreased in patients with advanced colorectal neoplasia compared to the controls. The ^{1}H-NMR was then correlated to other tests such as guaiac-fecal occult blood test and Wif-1 methylation test, where ^{1}H-NMR showed higher accuracy for predicting the advanced colorectal neoplasia. In another study by Lin *et al.* [100], ^{1}H-NMR was used to profile fecal metabolites of 68 CRC patients (stage I/II=20; stage III=25 and stage IV=23) and 32 healthy controls. The obtained metabolic fingerprint of healthy controls can be distinguished from CRC patients at various stages.

Urinary Bladder Cancer

Urinary bladder cancer (UBC) is one of the leading lethal cancers worldwide with an incidence of 68800 new cases per year in the USA alone [111]. The current standard approaches to diagnose UBC, including cystoscopy and histopathologic examination, are limited by the high cost and their invasive nature [112]. UBC is usually associated with high risk of recurrence, so UBC patients are monitored lifelong for recurrence through urine cytology, imaging and periodic cystoscopy. However, these monitoring techniques lack adequate sensitivity and specificity for clinical use [113]. Therefore, looking for sensitive and specific, cost-effective noninvasive tool to diagnose and monitor patients with UBC is of vital importance.

NMR spectroscopy was applied to identify novel and accurate biomarkers in serum and urine specimens that would help to early detect and distinguish UBC patients from healthy subjects and others with benign conditions. In the study of Cao *et al.* [111], ^{1}H NMR spectroscopic analysis was performed on serum samples obtained from UBC patients, healthy subjects and calculi patients with the similar clinic sign of hematuria as UBC patients. Serum metabolic profiles of UBC patients showed decreased levels of isoleucine/leucine, tyrosine, lactate, glycine, citrate, as well as elevated concentrations of lipids and glucose compared to both calculi and healthy subjects.

Urine is usually rich in tumor cells and metabolites that shed from the tumor mass in the bladder wall [113]. ^{1}H NMR-based metabolic analysis was applied by Srivastava *et al.* [114] to identify the metabolic signature of urine samples collected from 37 healthy individuals, 33 benign control patients (31 with urinary tract infection and 2 with bladder stone), and 33 patients with bladder cancer of different stages (8 Ta, 19 T1, 6 carcinoma in situ). The results showed that four metabolites were markedly altered, including citrate, hippurate, phenylalanine, and taurine as shown in Fig. (**13**). The concentration of taurine was significantly elevated in patients with bladder cancer relative to both healthy and benign control patients.

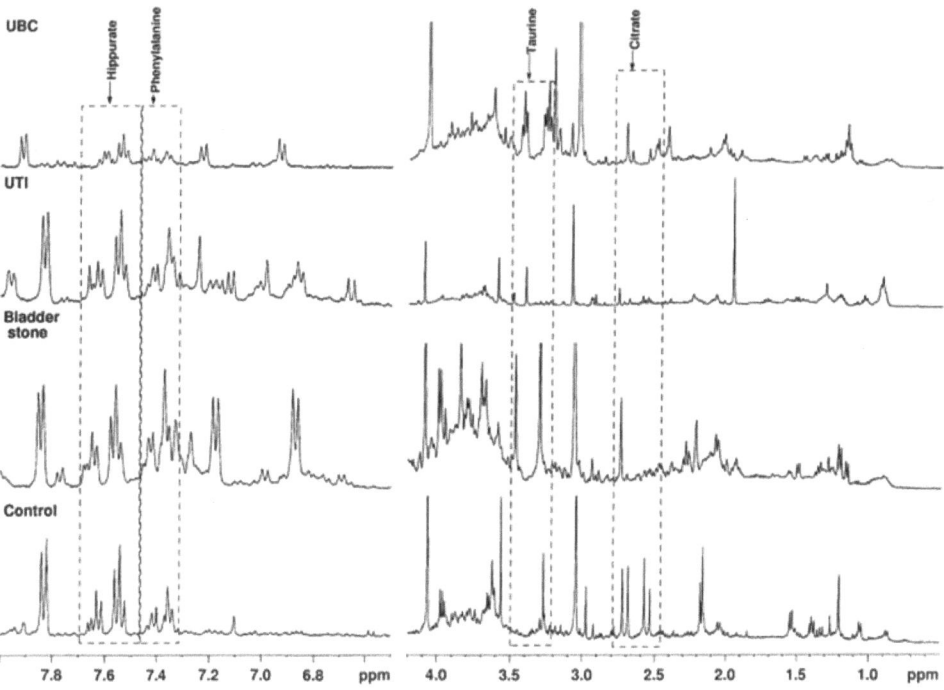

Fig. (13). The metabolic differences in spectra of healthy controls, controls with benign conditions (UTI and bladder stone) and bladder cancer patients have been depicted. The metabolic variations are quite prominent (Reprinted with permission from [114]).

Oral Cancer

Oral cancer (OC) is the sixth commonest cancer worldwide with high morbidity and mortality rates [115]. Patients diagnosed at late stages have 5-year survival rates less than 50%, whereas the early detection of OC would increase the chance of survival at 5 years up to 90% [116]. Absence of accurate biomarkers and lack of simple, noninvasive diagnostic approaches suitable for the screening of patients at early stages are responsible for the alarming mortality rates [117]. Therefore, the research on accurate and noninvasive screening test for detecting precancerous patients is urgently required.

Being a noninvasive approach, blood-based NMR spectroscopy was tested as a primary screening tool to identify novel metabolic biomarkers that would help the early detection of OC patients. In the study of Tiziani *et al.* [116], serum samples collected from 15 OC patients and 10 healthy controls were analyzed by ^{1}H NMR spectroscopy. The obtained metabolic signature for OC patients was differentiated from that of the healthy controls with sensitivity and specificity of more than

95%. Serum samples of cancer patients showed reduced levels of valine, ethanol, lactate, alanine, acetate, citrate, phenylalanine, tyrosine, methanol, formaldehyde, and formic acid as well as elevated concentrations of glucose, pyruvate, acetone, acetoacetate, 3-hydroxybutyrate and 2- hydroxybutyrate, choline, betaine, dimethylglycine, sarcosine, asparagine, and ornithine compared to the healthy controls. Subsequent analysis of metabolite profiles of serum samples from OC patients allowed differentiation between cancer patients with high- and low-stage diseases. Patients with late stage disease showed increased levels of 2-hydroxybutyrate, 3-hydroxybutyrate, acetone, acetate, acetoacetate, creatinine, asparagine, glucose, dimethylglycine, betaine, and choline as well as decreased concentrations of valine, lactate, alanine, pyruvate, lysine, creatine, acetyl--carnitine, and carnitine compared to patients with early stage disease. In another study of Bag *et al.* [118], ^{1}H and ^{13}C NMR analysis of serum samples of oral squamous cell carcinoma (OSCC) patients and control group revealed that choline was down-regulated, while its break-down product, trimethylamine N-oxide, was upregulated in OSCC patients compared to the healthy control. Moreover, no significant change was observed in lactate signal indicating that the well-known Warburg effect was not a prominent phenomenon in this type of tumor.

Oral leukoplakia (OLK) is characterized by white premalignant lesions on oral mucosa, which has a great tendency of malignant transformation to OSCC [119]. Therefore, the accurate detection and the adequate treatment of OLK patients would decrease the risk of OSCC development. ^{1}H NMR-based metabolomics was performed by Zhou *et al.* [120] to analyze plasma samples of patients with OSCC, patients with OLK, and the healthy control group. The markedly altered metabolites including choline, valine and lactate could discriminate between the 3 different groups as shown in Fig. (**14**). In another study of Gupta *et al.* [115], ^{1}H NMR spectroscopy was performed to analyze serum samples obtained from 100 OLK patients, 100 OSCC patients, and 75 healthy subjects. The results showed that four metabolites (glutamine, propionate, acetone, and choline) were able to distinguish 93.5% of OSCC patients from the healthy control with high sensitivity and specificity. Similarly, four biomarkers (glutamine, acetone, acetate, and choline) were able to differentiate between OSCC and OLK patients.

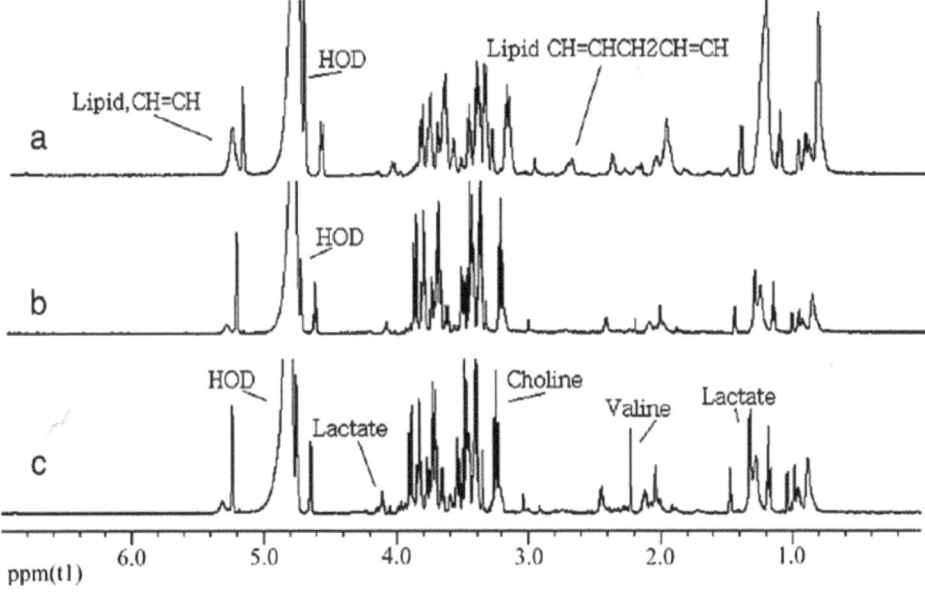

Fig. (14). The ^{1}H NMR spectra of the plasma from the OSCC patients, the OLK patients and the healthy controls; "a" showed the typical spectra of the plasma from the healthy control; "b" showed the typical spectra from the OLK patient with the pathological grade of moderate hyperplasia; "c" showed the typical spectra from the OSCC patient at stage T2 and pathological grade I (Reprinted with permission from [120]).

PERSPECTIVE AND CONCLUDING REMARKS

Although NMR spectroscopy is a very powerful tool in proteomics and metabolomics, its limited sensitivity is a major drawback. Substantial effort was exerted to increase the sensitivity of NMR using microprobes [121], high magnetic field strength [122], cryogenic probes [123], dynamic nuclear polarization [124] and chemical derivatization [125]. Increasing NMR sensitivity will raise another problem, because the NMR chart will be more complicated. The current NMR instruments can detect around 100 metabolites out of several hundreds. If the sensitivity could be increased, more metabolites would be sensed and more complicated charts will be obtained. Data interpretation in NMR is based on databases with the aid of multivariate analysis. More elaborated databases and mathematical treatments are expected in the future to cope with the progressive increase in sensitivity. Automation of NMR is another research trend in this field. Flow injection NMR allows the analysis of several hundreds of samples per day with less operator intervention, which saves time and effort and improves precision and accuracy.

Because only microliters of the sample are required for NMR screening, microextraction could be a good choice for sample preparation [126].

Microextraction differs from conventional extraction in the volume used for extraction, the amount retrieved of the analyte and the concentration. microextraction requires only a few microliters of extracting agent (solid or liquid), which decreases organic solvent consumption and protects the environment [127, 128]. Although the amount recovered of the analyte in microextraction is small, compared with conventional extraction, this small amount is highly concentrated due to the small volume of the extractant, which makes the sample highly enriched. More applications of microextraction as a sample preparation method before NMR are expected in the future.

CONSENT FOR PUBLICATION

Not applicable.

CONFLICT OF INTEREST

The authors confirm that this chapter contents have no conflict of interest.

ACKNOWLEDGEMENTS

Declared none.

REFERENCES

[1] Zia, K.; Siddiqui, T.; Ali, S.; Farooq, I.; Zafar, M.S.; Khurshid, Z. Nuclear Magnetic Resonance Spectroscopy for Medical and Dental Applications: A Comprehensive Review. *Eur. J. Dent.*, **2019**, *13*(1), 124-128.
[http://dx.doi.org/10.1055/s-0039-1688654] [PMID: 31170770]

[2] Bloch, F. Nuclear Induction. *Physica,* **1946**, *17*(3–4), 272-281.

[3] Cox, I.J.; Sharif, A.; Cobbold, J.F.L.; Thomas, H.C.; Taylor-Robinson, S.D. Current and future applications of *in vitro* magnetic resonance spectroscopy in hepatobiliary disease. *World J. Gastroenterol.,* **2006**, *12*(30), 4773-4783.
[http://dx.doi.org/10.3748/wjg.v12.i30.4773] [PMID: 16937457]

[4] Gika, H.G.; Theodoridis, G.A.; Plumb, R.S.; Wilson, I.D. Current practice of liquid chromatography-mass spectrometry in metabolomics and metabonomics. *J. Pharm. Biomed. Anal.,* **2014**, *87*, 12-25.
[http://dx.doi.org/10.1016/j.jpba.2013.06.032] [PMID: 23916607]

[5] Tognarelli, J.M.; Dawood, M.; Shariff, M.I.F.; Grover, V.P.B.; Crossey, M.M.E.; Cox, I.J.; Taylor-Robinson, S.D.; McPhail, M.J.W. Magnetic resonance spectroscopy: Principles and techniques: Lessons for clinicians. *J. Clin. Exp. Hepatol.,* **2015**, *5*(4), 320-328.
[http://dx.doi.org/10.1016/j.jceh.2015.10.006] [PMID: 26900274]

[6] Zhang, A.H.; Sun, H.; Qiu, S.; Wang, X.J. NMR-based metabolomics coupled with pattern recognition methods in biomarker discovery and disease diagnosis. *Magn. Reson. Chem.,* **2013**, *51*(9), 549-556.
[http://dx.doi.org/10.1002/mrc.3985] [PMID: 23828598]

[7] Capati, A.; Ijare, O. B.; Bezabeh, T. Diagnostic Applications of Nuclear Magnetic 750 Resonance – Based Urinary Metabolomics. *Magn. Reson. Insights,* **2017**, *10*, 1178623X17694346..

[8] Tietze, A.; Choi, C.; Mickey, B.; Maher, E.A.; Parm Ulhøi, B.; Sangill, R.; Lassen-Ramshad, Y.; Lukacova, S.; Østergaard, L.; von Oettingen, G. Noninvasive assessment of isocitrate dehydrogenase

mutation status in cerebral gliomas by magnetic resonance spectroscopy in a clinical setting. *J. Neurosurg.*, **2018**, *128*(2), 391-398.
[http://dx.doi.org/10.3171/2016.10.JNS161793] [PMID: 28298040]

[9] Majós, C.; Alonso, J.; Aguilera, C.; Serrallonga, M.; Pérez-Martín, J.; Acebes, J.J.; Arús, C.; Gili, J. Proton magnetic resonance spectroscopy ((1)H MRS) of human brain tumours: assessment of differences between tumour types and its applicability in brain tumour categorization. *Eur. Radiol.*, **2003**, *13*(3), 582-591.
[http://dx.doi.org/10.1007/s00330-002-1547-3] [PMID: 12594562]

[10] Emwas, A.H.; Luchinat, C.; Turano, P.; Tenori, L.; Roy, R.; Salek, R.M.; Ryan, D.; Merzaban, J.S.; Kaddurah-Daouk, R.; Zeri, A.C.; Nagana Gowda, G.A.; Raftery, D.; Wang, Y.; Brennan, L.; Wishart, D.S. Standardizing the experimental conditions for using urine in NMR-based metabolomic studies with a particular focus on diagnostic studies: a review. *Metabolomics*, **2015**, *11*(4), 872-894.
[http://dx.doi.org/10.1007/s11306-014-0746-7] [PMID: 26109927]

[11] Khan, S.A.; Cox, I.J.; Hamilton, G.; Thomas, H.C.; Taylor-Robinson, S.D. In vivo and *in vitro* nuclear magnetic resonance spectroscopy as a tool for investigating hepatobiliary disease: a review of H and P MRS applications. *Liver Int.*, **2005**, *25*(2), 273-281.
[http://dx.doi.org/10.1111/j.1478-3231.2005.01090.x] [PMID: 15780050]

[12] Chatham, J.C.; Blackband, S.J. Nuclear magnetic resonance spectroscopy and imaging in animal research. *ILAR J.*, **2001**, *42*(3), 189-208.
[http://dx.doi.org/10.1093/ilar.42.3.189] [PMID: 11406719]

[13] Geva, T. Magnetic resonance imaging: historical perspective. *J. Cardiovasc. Magn. Reson.*, **2006**, *8*(4), 573-580.
[http://dx.doi.org/10.1080/10976640600755302] [PMID: 16869310]

[14] Pavia, D.L.; Lampman, G.M.; Kriz, G.S.; Vyvyan, J. *Introduction to Spectroscopy*; Cengage Learning: Boston, Massachusetts, **2008**.

[15] Skoog, D.A.; Holler, F.J.; Crouch, S.R. *Principles of Instrumental Analysis*; Thomson Brooks/Cole: Philadelphia, **2007**.

[16] Giraudeau, P.; Frydman, L. Ultrafast 2D NMR: an emerging tool in analytical spectroscopy. *Annu. Rev. Anal. Chem. (Palo Alto, Calif.)*, **2014**, *7*(1), 129-161.
[http://dx.doi.org/10.1146/annurev-anchem-071213-020208] [PMID: 25014342]

[17] Brierley, B.; Shaw, P.; David, A.S. The human amygdala: a systematic review and meta-analysis of volumetric magnetic resonance imaging. *Brain Res. Brain Res. Rev.*, **2002**, *39*(1), 84-105.
[http://dx.doi.org/10.1016/S0165-0173(02)00160-1] [PMID: 12086710]

[18] Krägeloh-Mann, I.; Horber, V. The role of magnetic resonance imaging in elucidating the pathogenesis of cerebral palsy: a systematic review. *Dev. Med. Child Neurol.*, **2007**, *49*(2), 144-151.
[http://dx.doi.org/10.1111/j.1469-8749.2007.00144.x] [PMID: 17254004]

[19] Vijayalaxmi, ; Fatahi, M.; Speck, O. Magnetic resonance imaging (MRI): A review of genetic damage investigations. *Mutat. Res. Rev. Mutat. Res.*, **2015**, *764*, 51-63.
[http://dx.doi.org/10.1016/j.mrrev.2015.02.002] [PMID: 26041266]

[20] Boesch, C. Nobel Prizes for nuclear magnetic resonance: 2003 and historical perspectives. *J. Magn. Reson. Imaging*, **2004**, *20*(2), 177-179.
[http://dx.doi.org/10.1002/jmri.20120] [PMID: 15269938]

[21] Barding, G.A., Jr; Salditos, R.; Larive, C.K. Quantitative NMR for bioanalysis and metabolomics. *Anal. Bioanal. Chem.*, **2012**, *404*(4), 1165-1179.
[http://dx.doi.org/10.1007/s00216-012-6188-z] [PMID: 22766756]

[22] Silva Elipe, M.V. Advantages and Disadvantages of Nuclear Magnetic Resonance Spectroscopy as a Hyphenated Technique. *Anal. Chim. Acta*, **2003**, *497*(1–2), 1-25.
[http://dx.doi.org/10.1016/j.aca.2003.08.048]

[23] Emwas, A.H.M.; Salek, R.M.; Griffin, J.L.; Merzaban, J. NMR-Based Metabolomics in Human Disease Diagnosis: Applications, Limitations, and Recommendations. *Metabolomics*, **2013**, *9*(5), 1048-1072.
[http://dx.doi.org/10.1007/s11306-013-0524-y]

[24] Emwas, A.H.; Luchinat, C.; Turano, P.; Tenori, L.; Roy, R.; Salek, R.M.; Ryan, D.; Merzaban, J.S.; Kaddurah-Daouk, R.; Zeri, A.C. Standardizing the Experimental Conditions for Using Urine in NMR-Based Metabolomic Studies with a Particular Focus on Diagnostic Studies: A Review. *Metabolomics. Springer New York*, **2015**, *LLC*, 872-894.
[http://dx.doi.org/10.1007/s11306-014-0746-7]

[25] Barton, R.H.; Waterman, D.; Bonner, F.W.; Holmes, E.; Clarke, R.; Nicholson, J.K.; Lindon, J.C. The influence of EDTA and citrate anticoagulant addition to human plasma on information recovery from NMR-based metabolic profiling studies. *Mol. Biosyst.*, **2010**, *6*(1), 215-224.
[http://dx.doi.org/10.1039/b907021d] [PMID: 20024083]

[26] Pertinhez, T.A.; Casali, E.; Lindner, L.; Spisni, A.; Baricchi, R.; Berni, P. Biochemical assessment of red blood cells during storage by (1)H nuclear magnetic resonance spectroscopy. Identification of a biomarker of their level of protection against oxidative stress. *Blood Transfus.*, **2014**, *12*(4), 548-556.
[PMID: 24960643]

[27] Budde, K.; Gök, Ö.N.; Pietzner, M.; Meisinger, C.; Leitzmann, M.; Nauck, M.; Köttgen, A.; Friedrich, N. Quality assurance in the pre-analytical phase of human urine samples by (1)H NMR spectroscopy. *Arch. Biochem. Biophys.*, **2016**, *589*, 10-17.
[http://dx.doi.org/10.1016/j.abb.2015.07.016] [PMID: 26264917]

[28] Chen, W.; Lou, H.; Zhang, H.; Nie, X.; Lan, W.; Yang, Y.; Xiang, Y.; Qi, J.; Lei, H.; Tang, H.; Chen, F.; Deng, F. Grade classification of neuroepithelial tumors using high-resolution magic-angle spinning proton nuclear magnetic resonance spectroscopy and pattern recognition. *Sci. China Life Sci.*, **2011**, *54*(7), 606-616.
[http://dx.doi.org/10.1007/s11427-011-4193-7] [PMID: 21748584]

[29] Cuellar-Baena, S.; Morales, J.M.; Martinetto, H.; Calvar, J.; Sevlever, G.; Castellano, G.; Cerdá-Nicolás, M.; Celda, B.; Monleon, D. Comparative metabolic profiling of paediatric ependymoma, medulloblastoma and pilocytic astrocytoma. *Int. J. Mol. Med.*, **2010**, *26*(6), 941-948.
[PMID: 21042791]

[30] Sjøbakk, T.E.; Johansen, R.; Bathen, T.F.; Sonnewald, U.; Juul, R.; Torp, S.H.; Lundgren, S.; Gribbestad, I.S. Characterization of brain metastases using high-resolution magic angle spinning MRS. *NMR Biomed.*, **2008**, *21*(2), 175-185.
[http://dx.doi.org/10.1002/nbm.1180] [PMID: 17542042]

[31] Malet-Martino, M.; Holzgrabe, U. NMR techniques in biomedical and pharmaceutical analysis. *J. Pharm. Biomed. Anal.*, **2011**, *55*(1), 1-15.
[http://dx.doi.org/10.1016/j.jpba.2010.12.023] [PMID: 21237608]

[32] Lehnhardt, F.G.; Bock, C.; Röhn, G.; Ernestus, R.I.; Hoehn, M. Metabolic differences between primary and recurrent human brain tumors: a ^{1}H NMR spectroscopic investigation. *NMR Biomed.*, **2005**, *18*(6), 371-382.
[http://dx.doi.org/10.1002/nbm.968] [PMID: 15959923]

[33] Erb, G.; Elbayed, K.; Piotto, M.; Raya, J.; Neuville, A.; Mohr, M.; Maitrot, D.; Kehrli, P.; Namer, I.J. Toward improved grading of malignancy in oligodendrogliomas using metabolomics. *Magn. Reson. Med.*, **2008**, *59*(5), 959-965.
[http://dx.doi.org/10.1002/mrm.21486] [PMID: 18429037]

[34] Bathen, T.F.; Sitter, B.; Sjøbakk, T.E.; Tessem, M.B.; Gribbestad, I.S. Magnetic resonance metabolomics of intact tissue: a biotechnological tool in cancer diagnostics and treatment evaluation. *Cancer Res.*, **2010**, *70*(17), 6692-6696.
[http://dx.doi.org/10.1158/0008-5472.CAN-10-0437] [PMID: 20699363]

[35] Kelimu, A.; Xie, R.; Zhang, K.; Zhuang, Z.; Mamtimin, B.; Sheyhidin, I. Metabonomic signature analysis in plasma samples of glioma patients based on (^{1}H-nuclear magnetic resonance spectroscopy. *Neurol. India,* **2016**, *64*(2), 246-251.
[http://dx.doi.org/10.4103/0028-3886.177606] [PMID: 26954801]

[36] Wright, A.J.; Fellows, G.A.; Griffiths, J.R.; Wilson, M.; Bell, B.A.; Howe, F.A. *Ex-vivo* HRMAS of adult brain tumours: metabolite quantification and assignment of tumour biomarkers. *Mol. Cancer,* **2010**, *9*, 66.
[http://dx.doi.org/10.1186/1476-4598-9-66] [PMID: 20331867]

[37] Vettukattil, R.; Gulati, M.; Sjøbakk, T.E.; Jakola, A.S.; Kvernmo, N.A.M.; Torp, S.H.; Bathen, T.F.; Gulati, S.; Gribbestad, I.S. Differentiating diffuse World Health Organization grade II and IV astrocytomas with *ex vivo* magnetic resonance spectroscopy. *Neurosurgery,* **2013**, *72*(2), 186-195.
[http://dx.doi.org/10.1227/NEU.0b013e31827b9c57] [PMID: 23147779]

[38] Constantin, A.; Elkhaled, A.; Jalbert, L.; Srinivasan, R.; Cha, S.; Chang, S.M.; Bajcsy, R.; Nelson, S.J. Identifying malignant transformations in recurrent low grade gliomas using high resolution magic angle spinning spectroscopy. *Artif. Intell. Med.,* **2012**, *55*(1), 61-70.
[http://dx.doi.org/10.1016/j.artmed.2012.01.002] [PMID: 22387185]

[39] Elkhaled, A.; Jalbert, L.; Constantin, A.; Yoshihara, H.A.I.; Phillips, J.J.; Molinaro, A.M.; Chang, S.M.; Nelson, S.J. Characterization of metabolites in infiltrating gliomas using ex vivo ^{1}H high-resolution magic angle spinning spectroscopy. *NMR Biomed.,* **2014**, *27*(5), 578-593.
[http://dx.doi.org/10.1002/nbm.3097] [PMID: 24596146]

[40] Bademler, S.; Ucuncu, M.Z.; Tilgen Vatansever, C.; Serilmez, M.; Ertin, H.; Karanlık, H. Diagnostic and Prognostic Significance of Carboxypeptidase A4 (CPA4) in Breast Cancer. *Biomolecules,* **2019**, *9*(3), 103.
[http://dx.doi.org/10.3390/biom9030103] [PMID: 30875843]

[41] Rueda, O.M.; Sammut, S-J.; Seoane, J.A.; Chin, S-F.; Caswell-Jin, J.L.; Callari, M.; Batra, R.; Pereira, B.; Bruna, A.; Ali, H.R.; Provenzano, E.; Liu, B.; Parisien, M.; Gillett, C.; McKinney, S.; Green, A.R.; Murphy, L.; Purushotham, A.; Ellis, I.O.; Pharoah, P.D.; Rueda, C.; Aparicio, S.; Caldas, C.; Curtis, C. Dynamics of breast-cancer relapse reveal late-recurring ER-positive genomic subgroups. *Nature,* **2019**, *567*(7748), 399-404.
[http://dx.doi.org/10.1038/s41586-019-1007-8] [PMID: 30867590]

[42] Vermeersch, K.A.; Styczynski, M.P. Applications of metabolomics in cancer research. *J. Carcinog.,* **2013**, *12*(1), 9.
[http://dx.doi.org/10.4103/1477-3163.113622] [PMID: 23858297]

[43] Giskeødegård, G.F.; Grinde, M.T.; Sitter, B.; Axelson, D.E.; Lundgren, S.; Fjøsne, H.E.; Dahl, S.; Gribbestad, I.S.; Bathen, T.F. Multivariate modeling and prediction of breast cancer prognostic factors using MR metabolomics. *J. Proteome Res.,* **2010**, *9*(2), 972-979.
[http://dx.doi.org/10.1021/pr9008783] [PMID: 19994911]

[44] Sitter, B.; Lundgren, S.; Bathen, T.F.; Halgunset, J.; Fjosne, H.E.; Gribbestad, I.S. Comparison of HR MAS MR spectroscopic profiles of breast cancer tissue with clinical parameters. *NMR Biomed.,* **2006**, *19*(1), 30-40.
[http://dx.doi.org/10.1002/nbm.992] [PMID: 16229059]

[45] Gu, H.; Pan, Z.; Xi, B.; Asiago, V.; Musselman, B.; Raftery, D. Principal component directed partial least squares analysis for combining nuclear magnetic resonance and mass spectrometry data in metabolomics: application to the detection of breast cancer. *Anal. Chim. Acta,* **2011**, *686*(1-2), 57-63.
[http://dx.doi.org/10.1016/j.aca.2010.11.040] [PMID: 21237308]

[46] Slupsky, C.M.; Steed, H.; Wells, T.H.; Dabbs, K.; Schepansky, A.; Capstick, V.; Faught, W.; Sawyer, M.B. Urine metabolite analysis offers potential early diagnosis of ovarian and breast cancers. *Clin. Cancer Res.,* **2010**, *16*(23), 5835-5841.
[http://dx.doi.org/10.1158/1078-0432.CCR-10-1434] [PMID: 20956617]

[47] Singh, A.; Sharma, R.K.; Chagtoo, M.; Agarwal, G.; George, N.; Sinha, N.; Godbole, M.M. ^{1}H NMR Metabolomics Reveals Association of High Expression of Inositol 1, 4, 5 Trisphosphate Receptor and Metabolites in Breast Cancer Patients. *PLoS One,* **2017**, *12*(1): e0169330.
[http://dx.doi.org/10.1371/journal.pone.0169330] [PMID: 28072864]

[48] Jobard, E.; Pontoizeau, C.; Blaise, B.J.; Bachelot, T.; Elena-Herrmann, B.; Trédan, O. A serum nuclear magnetic resonance-based metabolomic signature of advanced metastatic human breast cancer. *Cancer Lett.,* **2014**, *343*(1), 33-41.
[http://dx.doi.org/10.1016/j.canlet.2013.09.011] [PMID: 24041867]

[49] Hart, C.D.; Vignoli, A.; Tenori, L.; Uy, G.L.; Van To, T.; Adebamowo, C.; Hossain, S.M.; Biganzoli, L.; Risi, E.; Love, R.R.; Luchinat, C.; Di Leo, A. Serum metabolomic profiles identify er-positive early breast cancer patients at increased risk of disease recurrence in a multicenter population. *Clin. Cancer Res.,* **2017**, *23*(6), 1422-1431.
[http://dx.doi.org/10.1158/1078-0432.CCR-16-1153] [PMID: 28082280]

[50] Tenori, L.; Oakman, C.; Claudino, W.M.; Bernini, P.; Cappadona, S.; Nepi, S.; Biganzoli, L.; Arbushites, M.C.; Luchinat, C.; Bertini, I.; Di Leo, A. Exploration of serum metabolomic profiles and outcomes in women with metastatic breast cancer: a pilot study. *Mol. Oncol.,* **2012**, *6*(4), 437-444.
[http://dx.doi.org/10.1016/j.molonc.2012.05.003] [PMID: 22687601]

[51] Bergstrom, J.; Shih, I-M.; Fader, A.N. Updates on Rare Epithelial Ovarian Carcinoma. In: *Translational Advances in Gynecologic Cancers*; Birrer, M.J., Ceppi, L. Eds.; Academic Press: Cambridge, Massachusetts, **2017**; pp. 181-193.
[http://dx.doi.org/10.1016/B978-0-12-803741-6.00010-0]

[52] Doubeni, C.A.; Doubeni, A.R.B.; Myers, A.E. Diagnosis and management of ovarian cancer. *Am. Fam. Physician,* **2016**, *93*(11), 937-944.
[PMID: 27281838]

[53] Torre, L.A.; Trabert, B.; DeSantis, C.E.; Miller, K.D.; Samimi, G.; Runowicz, C.D.; Gaudet, M.M.; Jemal, A.; Siegel, R.L. Ovarian cancer statistics, 2018. *CA Cancer J. Clin.,* **2018**, *68*(4), 284-296.
[http://dx.doi.org/10.3322/caac.21456] [PMID: 29809280]

[54] Zhang, T.; Wu, X.; Ke, C.; Yin, M.; Li, Z.; Fan, L.; Zhang, W.; Zhang, H.; Zhao, F.; Zhou, X.; Lou, G.; Li, K. Identification of potential biomarkers for ovarian cancer by urinary metabolomic profiling. *J. Proteome Res.,* **2013**, *12*(1), 505-512.
[http://dx.doi.org/10.1021/pr3009572] [PMID: 23163809]

[55] Boss, E.A.; Moolenaar, S.H.; Massuger, L.F.A.G.; Boonstra, H.; Engelke, U.F.H.; de Jong, J.G.N.; Wevers, R.A. High-resolution proton nuclear magnetic resonance spectroscopy of ovarian cyst fluid. *NMR Biomed.,* **2000**, *13*(5), 297-305.
[http://dx.doi.org/10.1002/1099-1492(200008)13:5<297::AID-NBM648>3.0.CO;2-I] [PMID: 10960920]

[56] Odunsi, K.; Wollman, R.M.; Ambrosone, C.B.; Hutson, A.; McCann, S.E.; Tammela, J.; Geisler, J.P.; Miller, G.; Sellers, T.; Cliby, W.; Qian, F.; Keitz, B.; Intengan, M.; Lele, S.; Alderfer, J.L. Detection of epithelial ovarian cancer using 1H-NMR-based metabonomics. *Int. J. Cancer,* **2005**, *113*(5), 782-788.
[http://dx.doi.org/10.1002/ijc.20651] [PMID: 15499633]

[57] Garcia, E.; Andrews, C.; Hua, J.; Kim, H.L.; Sukumaran, D.K.; Szyperski, T.; Odunsi, K. Diagnosis of early stage ovarian cancer by ^{1}H NMR metabonomics of serum explored by use of a microflow NMR probe. *J. Proteome Res.,* **2011**, *10*(4), 1765-1771.
[http://dx.doi.org/10.1021/pr101050d] [PMID: 21218854]

[58] Alix-Panabières, C.; Mader, S.; Pantel, K. Epithelial-mesenchymal plasticity in circulating tumor cells. *J. Mol. Med. (Berl.),* **2017**, *95*(2), 133-142.
[http://dx.doi.org/10.1007/s00109-016-1500-6] [PMID: 28013389]

[59] Ghazani, A.A.; Castro, C.M.; Gorbatov, R.; Lee, H.; Weissleder, R. Sensitive and direct detection of

circulating tumor cells by multimarker μ-nuclear magnetic resonance. *Neoplasia,* **2012**, *14*(5), 388-395.
[http://dx.doi.org/10.1596/neo.12696] [PMID: 22745585]

[60] Bharti, S.K.; Wildes, F.; Hung, C.F.; Wu, T.C.; Bhujwalla, Z.M.; Penet, M.F. Metabolomic characterization of experimental ovarian cancer ascitic fluid. *Metabolomics,* **2017**, *13*(10), 1-11.
[http://dx.doi.org/10.1007/s11306-017-1254-3] [PMID: 29430218]

[61] Webb, P.M.; Na, R.; Weiderpass, E.; Adami, H.O.; Anderson, K.E.; Bertrand, K.A.; Botteri, E.; Brasky, T.M.; Brinton, L.A.; Chen, C.; Doherty, J.A.; Lu, L.; McCann, S.E.; Moysich, K.B.; Olson, S.; Petruzella, S.; Palmer, J.R.; Prizment, A.E.; Schairer, C.; Setiawan, V.W.; Spurdle, A.B.; Trabert, B.; Wentzensen, N.; Wilkens, L.; Yang, H.P.; Yu, H.; Risch, H.A.; Jordan, S.J. Use of aspirin, other nonsteroidal anti-inflammatory drugs and acetaminophen and risk of endometrial cancer: the Epidemiology of Endometrial Cancer Consortium. *Ann. Oncol.,* **2019**, *30*(2), 310-316.
[http://dx.doi.org/10.1093/annonc/mdy541] [PMID: 30566587]

[62] Bahado-Singh, R.O.; Lugade, A.; Field, J.; Al-Wahab, Z.; Han, B.; Mandal, R.; Bjorndahl, T.C.; Turkoglu, O.; Graham, S.F.; Wishart, D.; Odunsi, K. Metabolomic prediction of endometrial cancer. *Metabolomics,* **2017**, *14*(1), 6.
[http://dx.doi.org/10.1007/s11306-017-1290-z] [PMID: 30830361]

[63] Bray, F.; Ferlay, J.; Soerjomataram, I.; Siegel, R.L.; Torre, L.A.; Jemal, A. Global cancer statistics 2018: GLOBOCAN estimates of incidence and mortality worldwide for 36 cancers in 185 countries. *CA Cancer J. Clin.,* **2018**, *68*(6), 394-424.
[http://dx.doi.org/10.3322/caac.21492] [PMID: 30207593]

[64] Chiu, S.; Adcock, L. *Magnetic Resonance Imaging for Prostate Assessment: A Review of Clinical and Cost-Effectiveness*; Magn. Reson. Imaging Prostate Assess. A Rev. Clin. Cost-Effectiveness, **2018**, pp. 1-49.

[65] Kdadra, M.; Höckner, S.; Leung, H.; Kremer, W.; Schiffer, E. Metabolomics Biomarkers of Prostate Cancer: A Systematic Review. *Diagnostics (Basel),* **2019**, *9*(1), 21.

[66] Kasivisvanathan, V.; Rannikko, A.S.; Borghi, M.; Panebianco, V.; Mynderse, L.A.; Vaarala, M.H.; Briganti, A.; Budäus, L.; Hellawell, G.; Hindley, R.G.; Roobol, M.J.; Eggener, S.; Ghei, M.; Villers, A.; Bladou, F.; Villeirs, G.M.; Virdi, J.; Boxler, S.; Robert, G.; Singh, P.B.; Venderink, W.; Hadaschik, B.A.; Ruffion, A.; Hu, J.C.; Margolis, D.; Crouzet, S.; Klotz, L.; Taneja, S.S.; Pinto, P.; Gill, I.; Allen, C.; Giganti, F.; Freeman, A.; Morris, S.; Punwani, S.; Williams, N.R.; Brew-Graves, C.; Deeks, J.; Takwoingi, Y.; Emberton, M.; Moore, C.M. MRI-Targeted or Standard Biopsy for Prostate-Cancer Diagnosis. *N. Engl. J. Med.,* **2018**, *378*(19), 1767-1777.
[http://dx.doi.org/10.1056/NEJMoa1801993] [PMID: 29552975]

[67] Yang, B.; Liao, G.Q.; Wen, X.F.; Chen, W.H.; Cheng, S.; Stolzenburg, J.U.; Ganzer, R.; Neuhaus, J. Nuclear magnetic resonance spectroscopy as a new approach for improvement of early diagnosis and risk stratification of prostate cancer. *J. Zhejiang Univ. Sci. B,* **2017**, *18*(11), 921-933.
[http://dx.doi.org/10.1631/jzus.B1600441] [PMID: 29119730]

[68] Zaragozá, P.; Ruiz-Cerdá, J.L.; Quintás, G.; Gil, S.; Costero, A.M.; León, Z.; Vivancos, J.L.; Martínez-Máñez, R. Towards the potential use of (1)H NMR spectroscopy in urine samples for prostate cancer detection. *Analyst (Lond.),* **2014**, *139*(16), 3875-3878.
[http://dx.doi.org/10.1039/C4AN00690A] [PMID: 24989437]

[69] Pérez-Rambla, C.; Puchades-Carrasco, L.; García-Flores, M.; Rubio-Briones, J.; López-Guerrero, J.A.; Pineda-Lucena, A. Non-invasive urinary metabolomic profiling discriminates prostate cancer from benign prostatic hyperplasia. *Metabolomics,* **2017**, *13*(5), 52.
[http://dx.doi.org/10.1007/s11306-017-1194-y] [PMID: 28804274]

[70] Kumar, D.; Gupta, A.; Mandhani, A.; Sankhwar, S.N. Metabolomics-derived prostate cancer biomarkers: fact or fiction? *J. Proteome Res.,* **2015**, *14*(3), 1455-1464.
[http://dx.doi.org/10.1021/pr5011108] [PMID: 25609016]

[71] Kumar, D.; Gupta, A.; Mandhani, A.; Sankhwar, S.N. NMR spectroscopy of filtered serum of prostate cancer: A new frontier in metabolomics. *Prostate,* **2016**, *76*(12), 1106-1119.
[http://dx.doi.org/10.1002/pros.23198] [PMID: 27197810]

[72] Song, Z.; Wang, H.; Yin, X.; Deng, P.; Jiang, W. Application of NMR metabolomics to search for human disease biomarkers in blood. *Clin. Chem. Lab. Med.,* **2019**, *57*(4), 417-441.
[http://dx.doi.org/10.1515/cclm-2018-0380] [PMID: 30169327]

[73] Beger, R.D. A review of applications of metabolomics in cancer. *Metabolites,* **2013**, *3*(3), 552-574.
[http://dx.doi.org/10.3390/metabo3030552] [PMID: 24958139]

[74] Culig, Z. Distinguishing indolent from aggressive prostate cancer. *Recent Results Cancer Res.,* **2014**, *202*, 141-147.
[http://dx.doi.org/10.1007/978-3-642-45195-9_17] [PMID: 24531788]

[75] Giskeødegård, G.F.; Bertilsson, H.; Selnæs, K.M.; Wright, A.J.; Bathen, T.F.; Viset, T.; Halgunset, J.; Angelsen, A.; Gribbestad, I.S.; Tessem, M.B. Spermine and citrate as metabolic biomarkers for assessing prostate cancer aggressiveness. *PLoS One,* **2013**, *8*(4): e62375.
[http://dx.doi.org/10.1371/journal.pone.0062375] [PMID: 23626811]

[76] Vandergrift, L.A.; Decelle, E.A.; Kurth, J.; Wu, S.; Fuss, T.L.; DeFeo, E.M.; Halpern, E.F.; Taupitz, M.; McDougal, W.S.; Olumi, A.F.; Wu, C.L.; Cheng, L.L. Metabolomic Prediction of Human Prostate Cancer Aggressiveness: Magnetic Resonance Spectroscopy of Histologically Benign Tissue. *Sci. Rep.,* **2018**, *8*(1), 4997.
[http://dx.doi.org/10.1038/s41598-018-23177-w] [PMID: 29581441]

[77] Jordan, K.W.; Cheng, L.L. NMR-based metabolomics approach to target biomarkers for human prostate cancer. *Expert Rev. Proteomics,* **2007**, *4*(3), 389-400.
[http://dx.doi.org/10.1586/14789450.4.3.389] [PMID: 17552923]

[78] Kumar, V.; Dwivedi, D.K.; Jagannathan, N.R. High-resolution NMR spectroscopy of human body fluids and tissues in relation to prostate cancer. *NMR Biomed.,* **2014**, *27*(1), 80-89.
[http://dx.doi.org/10.1002/nbm.2979] [PMID: 23828638]

[79] Roberts, M.J.; Richards, R.S.; Chow, C.W.K.; Buck, M.; Yaxley, J.; Lavin, M.F.; Schirra, H.J.; Gardiner, R.A. Seminal plasma enables selection and monitoring of active surveillance candidates using nuclear magnetic resonance-based metabolomics: A preliminary investigation. *Prostate Int.,* **2017**, *5*(4), 149-157.
[http://dx.doi.org/10.1016/j.prnil.2017.03.005] [PMID: 29188202]

[80] Serkova, N.J.; Gamito, E.J.; Jones, R.H.; O'Donnell, C.; Brown, J.L.; Green, S.; Sullivan, H.; Hedlund, T.; Crawford, E.D. The metabolites citrate, myo-inositol, and spermine are potential age-independent markers of prostate cancer in human expressed prostatic secretions. *Prostate,* **2008**, *68*(6), 620-628.
[http://dx.doi.org/10.1002/pros.20727] [PMID: 18213632]

[81] Chen, W.; Lu, S.; Ou, J.; Wang, G.; Zu, Y.; Chen, F.; Bai, C. Metabonomic characteristics and biomarker research of human lung cancer tissues by HR^{1}H NMR spectroscopy. *Cancer Biomark.,* **2016**, *16*(4), 653-664.
[http://dx.doi.org/10.3233/CBM-160607] [PMID: 27002768]

[82] Rocha, C.M.; Carrola, J.; Barros, A.S.; Gil, A.M.; Goodfellow, B.J.; Carreira, I.M.; Bernardo, J.; Gomes, A.; Sousa, V.; Carvalho, L.; Duarte, I.F. Metabolic signatures of lung cancer in biofluids: NMR-based metabonomics of blood plasma. *J. Proteome Res.,* **2011**, *10*(9), 4314-4324.
[http://dx.doi.org/10.1021/pr200550p] [PMID: 21744875]

[83] Collins, L.G.; Haines, C.; Perkel, R.; Enck, R.E. Lung cancer: diagnosis and management. *Am. Fam. Physician,* **2007**, *75*(1), 56-63.
[PMID: 17225705]

[84] Carrola, J.; Rocha, C.M.; Barros, A.S.; Gil, A.M.; Goodfellow, B.J.; Carreira, I.M.; Bernardo, J.;

Gomes, A.; Sousa, V.; Carvalho, L.; Duarte, I.F. Metabolic signatures of lung cancer in biofluids: NMR-based metabonomics of urine. *J. Proteome Res.,* **2011**, *10*(1), 221-230.
[http://dx.doi.org/10.1021/pr100899x] [PMID: 21058631]

[85] Louis, E.; Adriaensens, P.; Guedens, W.; Bigirumurame, T.; Baeten, K.; Vanhove, K.; Vandeurzen, K.; Darquennes, K.; Vansteenkiste, J.; Dooms, C.; Shkedy, Z.; Mesotten, L.; Thomeer, M. Detection of Lung Cancer through Metabolic Changes Measured in Blood Plasma. *J. Thorac. Oncol.,* **2016**, *11*(4), 516-523.
[http://dx.doi.org/10.1016/j.jtho.2016.01.011] [PMID: 26949046]

[86] Rocha, C.M.; Barros, A.S.; Gil, A.M.; Goodfellow, B.J.; Humpfer, E.; Spraul, M.; Carreira, I.M.; Melo, J.B.; Bernardo, J.; Gomes, A.; Sousa, V.; Carvalho, L.; Duarte, I.F. Metabolic profiling of human lung cancer tissue by 1H high resolution magic angle spinning (HRMAS) NMR spectroscopy. *J. Proteome Res.,* **2010**, *9*(1), 319-332.
[http://dx.doi.org/10.1021/pr9006574] [PMID: 19908917]

[87] Chen, W.; Zu, Y.; Huang, Q.; Chen, F.; Wang, G.; Lan, W.; Bai, C.; Lu, S.; Yue, Y.; Deng, F. Study on metabonomic characteristics of human lung cancer using high resolution magic-angle spinning ^{1}H NMR spectroscopy and multivariate data analysis. *Magn. Reson. Med.,* **2011**, *66*(6), 1531-1540.
[http://dx.doi.org/10.1002/mrm.22957] [PMID: 21523825]

[88] Ahmed, N.; Bezabeh, T.; Ijare, O. B.; Myers, R.; Alomran, R.; Aliani, M.; Nugent, Z.; Banerji, S.; Kim, J.; Qing, G. Metabolic Signatures of Lung Cancer in Sputum and Exhaled Breath Condensate Detected by 1 H Magnetic Resonance Spectroscopy: A Feasibility Study. *Magn. Reson. Insights,* **2016**, *9*, MRI.S40864.

[89] Puchades-Carrasco, L.; Jantus-Lewintre, E.; Pérez-Rambla, C.; García-García, F.; Lucas, R.; Calabuig, S.; Blasco, A.; Dopazo, J.; Camps, C.; Pineda-Lucena, A. Serum metabolomic profiling facilitates the non-invasive identification of metabolic biomarkers associated with the onset and progression of non-small cell lung cancer. *Oncotarget,* **2016**, *7*(11), 12904-12916.
[http://dx.doi.org/10.18632/oncotarget.7354] [PMID: 26883203]

[90] Rocha, C.M.; Barros, A.S.; Goodfellow, B.J.; Carreira, I.M.; Gomes, A.; Sousa, V.; Bernardo, J.; Carvalho, L.; Gil, A.M.; Duarte, I.F. NMR metabolomics of human lung tumours reveals distinct metabolic signatures for adenocarcinoma and squamous cell carcinoma. *Carcinogenesis,* **2015**, *36*(1), 68-75.
[http://dx.doi.org/10.1093/carcin/bgu226] [PMID: 25368033]

[91] Chen, W.; Lu, S.; Wang, G.; Chen, F.; Bai, C. Staging research of human lung cancer tissues by high-resolution magic angle spinning proton nuclear magnetic resonance spectroscopy (HRMAS ^{1}H NMR) and multivariate data analysis. *Asia Pac. J. Clin. Oncol.,* **2017**, *13*(5), e232-e238.
[http://dx.doi.org/10.1111/ajco.12598] [PMID: 27670847]

[92] Hao, D.; Sarfaraz, M.O.; Farshidfar, F.; Bebb, D.G.; Lee, C.Y.; Card, C.M.; David, M.; Weljie, A.M. Temporal characterization of serum metabolite signatures in lung cancer patients undergoing treatment. *Metabolomics,* **2016**, *12*(3), 58.
[http://dx.doi.org/10.1007/s11306-016-0961-5] [PMID: 27073350]

[93] Deja, S.; Porebska, I.; Kowal, A.; Zabek, A.; Barg, W.; Pawelczyk, K.; Stanimirova, I.; Daszykowski, M.; Korzeniewska, A.; Jankowska, R.; Mlynarz, P. Metabolomics provide new insights on lung cancer staging and discrimination from chronic obstructive pulmonary disease. *J. Pharm. Biomed. Anal.,* **2014**, *100*, 369-380.
[http://dx.doi.org/10.1016/j.jpba.2014.08.020] [PMID: 25213261]

[94] Zhang, X.; Zhang, H.; Shen, B.; Sun, X.F. Chromogranin-A Expression as a Novel Biomarker for Early Diagnosis of Colon Cancer Patients. *Int. J. Mol. Sci.,* **2019**, *20*(12): E2919.
[http://dx.doi.org/10.3390/ijms20122919] [PMID: 31207989]

[95] Simon, K. Colorectal cancer development and advances in screening. *Clin. Interv. Aging,* **2016**, *11*, 967-976.
[http://dx.doi.org/10.2147/CIA.S109285] [PMID: 27486317]

[96] Ye, X.; Huai, J.; Ding, J. Diagnostic accuracy of fecal calprotectin for screening patients with colorectal cancer: A meta-analysis. *Turk. J. Gastroenterol.,* **2018**, *29*(4), 397-405.
[http://dx.doi.org/10.5152/tjg.2018.17606] [PMID: 30249553]

[97] Strul, H.; Arber, N. Screening techniques for prevention and early detection of colorectal cancer in the average-risk population. *Gastrointest. Cancer Res.,* **2007**, *1*(3), 98-106.
[PMID: 19262715]

[98] Bailey, J.R.; Aggarwal, A.; Imperiale, T.F. Colorectal Cancer Screening: Stool DNA and Other Noninvasive Modalities. *Gut Liver,* **2016**, *10*(2), 204-211.
[http://dx.doi.org/10.5009/gnl15420] [PMID: 26934885]

[99] Song, L-L.; Li, Y-M. Current noninvasive tests for colorectal cancer screening: An overview of colorectal cancer screening tests. *World J. Gastrointest. Oncol.,* **2016**, *8*(11), 793-800.
[http://dx.doi.org/10.4251/wjgo.v8.i11.793] [PMID: 27895817]

[100] Lin, Y.; Ma, C.; Liu, C.; Wang, Z.; Yang, J.; Liu, X.; Shen, Z.; Wu, R. NMR-based fecal metabolomics fingerprinting as predictors of earlier diagnosis in patients with colorectal cancer. *Oncotarget,* **2016**, *7*(20), 29454-29464.
[http://dx.doi.org/10.18632/oncotarget.8762] [PMID: 27107423]

[101] Zamani, Z.; Arjmand, M.; Vahabi, F.; Hosseini, S. M. E.; Fazeli, S. M.; Iravani, A.; Bayat, P.; Oghalayee, A.; Mehrabanfar, M.; Hosseini, R. H. A Metabolic Study 1059 on Colon Cancer Using 1H Nuclear Magnetic Resonance Spectroscopy. *Biochem. Res. Int.,* **2014**, *2014*.

[102] Wang, Z.; Lin, Y.; Liang, J.; Huang, Y.; Ma, C.; Liu, X.; Yang, J. NMR-based metabolomic techniques identify potential urinary biomarkers for early colorectal cancer detection. *Oncotarget,* **2017**, *8*(62), 105819-105831.
[http://dx.doi.org/10.18632/oncotarget.22402] [PMID: 29285295]

[103] Mirnezami, R.; Jiménez, B.; Li, J.V.; Kinross, J.M.; Veselkov, K.; Goldin, R.D.; Holmes, E.; Nicholson, J.K.; Darzi, A. Rapid diagnosis and staging of colorectal cancer *via* high-resolution magic angle spinning nuclear magnetic resonance (HR-MAS NMR) spectroscopy of intact tissue biopsies. *Ann. Surg.,* **2014**, *259*(6), 1138-1149.
[http://dx.doi.org/10.1097/SLA.0b013e31829d5c45] [PMID: 23860197]

[104] Bertini, I.; Cacciatore, S.; Jensen, B.V.; Schou, J.V.; Johansen, J.S.; Kruhøffer, M.; Luchinat, C.; Nielsen, D.L.; Turano, P. Metabolomic NMR fingerprinting to identify and predict survival of patients with metastatic colorectal cancer. *Cancer Res.,* **2012**, *72*(1), 356-364.
[http://dx.doi.org/10.1158/0008-5472.CAN-11-1543] [PMID: 22080567]

[105] Zhang, H.; Qiao, L.; Li, X.; Wan, Y.; Yang, L.; Wang, H. Tissue metabolic profiling of lymph node metastasis of colorectal cancer assessed by ^{1}H NMR. *Oncol. Rep.,* **2016**, *36*(6), 3436-3448.
[http://dx.doi.org/10.3892/or.2016.5175] [PMID: 27748865]

[106] Chan, E.C.; Koh, P.K.; Mal, M.; Cheah, P.Y.; Eu, K.W.; Backshall, A.; Cavill, R.; Nicholson, J.K.; Keun, H.C.; Keun, H.C. Metabolic profiling of human colorectal cancer using high-resolution magic angle spinning nuclear magnetic resonance (HR-MAS NMR) spectroscopy and gas chromatography mass spectrometry (GC/MS). *J. Proteome Res.,* **2009**, *8*(1), 352-361.
[http://dx.doi.org/10.1021/pr8006232] [PMID: 19063642]

[107] Jiménez, B.; Mirnezami, R.; Kinross, J.; Cloarec, O.; Keun, H.C.; Holmes, E.; Goldin, R.D.; Ziprin, P.; Darzi, A.; Nicholson, J.K. 1H HR-MAS NMR spectroscopy of tumor-induced local metabolic "field-effects" enables colorectal cancer staging and prognostication. *J. Proteome Res.,* **2013**, *12*(2), 959-968.
[http://dx.doi.org/10.1021/pr3010106] [PMID: 23240862]

[108] Vahabi, F.; Sadeghi, S.; Arjmand, M.; Mirkhani, F.; Hosseini, E.; Mehrabanfar, M.; Hajhosseini, R.; Iravani, A.; Bayat, P.; Zamani, Z. Staging of colorectal cancer using serum metabolomics with ^{1}HNMR Spectroscopy. *Iran. J. Basic Med. Sci.,* **2017**, *20*(7), 835-840.
[PMID: 28852450]

[109] Goedert, J.J.; Sampson, J.N.; Moore, S.C.; Xiao, Q.; Xiong, X.; Hayes, R.B.; Ahn, J.; Shi, J.; Sinha, R. Fecal metabolomics: assay performance and association with colorectal cancer. *Carcinogenesis*, **2014**, *35*(9), 2089-2096.
[http://dx.doi.org/10.1093/carcin/bgu131] [PMID: 25037050]

[110] Amiot, A.; Dona, A.C.; Wijeyesekera, A.; Tournigand, C.; Baumgaertner, I.; Lebaleur, Y.; Sobhani, I.; Holmes, E. (^{1}H NMR Spectroscopy of Fecal Extracts Enables Detection of Advanced Colorectal Neoplasia. *J. Proteome Res.*, **2015**, *14*(9), 3871-3881.
[http://dx.doi.org/10.1021/acs.jproteome.5b00277] [PMID: 26211820]

[111] Cao, M.; Zhao, L.; Chen, H.; Xue, W.; Lin, D. NMR-based metabolomic analysis of human bladder cancer. *Anal. Sci.*, **2012**, *28*(5), 451-456.
[http://dx.doi.org/10.2116/analsci.28.451] [PMID: 22687923]

[112] Hyndman, M.E.; Mullins, J.K.; Bivalacqua, T.J. Metabolomics and bladder cancer. *Urol. Oncol.*, **2011**, *29*(5), 558-561.
[http://dx.doi.org/10.1016/j.urolonc.2011.05.014] [PMID: 21930087]

[113] Zhang, J.; Wei, S.; Liu, L.; Nagana Gowda, G.A.; Bonney, P.; Stewart, J.; Knapp, D.W.; Raftery, D. NMR-based metabolomics study of canine bladder cancer. *Biochim. Biophys. Acta*, **2012**, *1822*(11), 1807-1814.
[http://dx.doi.org/10.1016/j.bbadis.2012.08.001] [PMID: 22967815]

[114] Srivastava, S.; Roy, R.; Singh, S.; Kumar, P.; Dalela, D.; Sankhwar, S.N.; Goel, A.; Sonkar, A.A. Taurine - a possible fingerprint biomarker in non-muscle invasive bladder cancer: A pilot study by ^{1}H NMR spectroscopy. *Cancer Biomark.*, **2010**, *6*(1), 11-20.
[http://dx.doi.org/10.3233/CBM-2009-0115] [PMID: 20164538]

[115] Gupta, A.; Gupta, S.; Mahdi, A.A. ^{1}H NMR-derived serum metabolomics of leukoplakia and squamous cell carcinoma. *Clin. Chim. Acta*, **2015**, *441*, 47-55.
[http://dx.doi.org/10.1016/j.cca.2014.12.003] [PMID: 25499120]

[116] Tiziani, S.; Lopes, V.; Günther, U.L. Early stage diagnosis of oral cancer using ^{1}H NMR-based metabolomics. *Neoplasia*, **2009**, *11*(3), 269-276, 4p, 269.
[http://dx.doi.org/10.1593/neo.81396] [PMID: 19242608]

[117] Mikkonen, J.J.W.; Singh, S.P.; Herrala, M.; Lappalainen, R.; Myllymaa, S.; Kullaa, A.M. Salivary metabolomics in the diagnosis of oral cancer and periodontal diseases. *J. Periodontal Res.*, **2016**, *51*(4), 431-437.
[http://dx.doi.org/10.1111/jre.12327] [PMID: 26446036]

[118] Bag, S.; Banerjee, D.R.; Basak, A.; Das, A.K.; Pal, M.; Banerjee, R.; Paul, R.R.; Chatterjee, J. NMR ((1)H and (13)C) based signatures of abnormal choline metabolism in oral squamous cell carcinoma with no prominent Warburg effect. *Biochem. Biophys. Res. Commun.*, **2015**, *459*(4), 574-578.
[http://dx.doi.org/10.1016/j.bbrc.2015.02.149] [PMID: 25769954]

[119] Bewley, A.F.; Farwell, D.G. Oral leukoplakia and oral cavity squamous cell carcinoma. *Clin. Dermatol.*, **2017**, *35*(5), 461-467.
[http://dx.doi.org/10.1016/j.clindermatol.2017.06.008] [PMID: 28916027]

[120] Zhou, J.; Xu, B.; Huang, J.; Jia, X.; Xue, J.; Shi, X.; Xiao, L.; Li, W. ^{1}H NMR-based metabonomic and pattern recognition analysis for detection of oral squamous cell carcinoma. *Clin. Chim. Acta*, **2009**, *401*(1-2), 8-13.
[http://dx.doi.org/10.1016/j.cca.2008.10.030] [PMID: 19056370]

[121] Dechow, J.; Forchel, A.; Lanz, T.; Haase, A. Fabrication of NMR - Microsensors for Nanoliter Sample Volumes. *Microelectron. Eng.*, **2000**, *53*(1), 517-519.
[http://dx.doi.org/10.1016/S0167-9317(00)00368-3]

[122] Gruetter, R.; Weisdorf, S.A.; Rajanayagan, V.; Terpstra, M.; Merkle, H.; Truwit, C.L.; Garwood, M.; Nyberg, S.L.; Uğurbil, K. Resolution improvements in *in vivo*^{1}H NMR spectra with increased

magnetic field strength. *J. Magn. Reson.,* **1998**, *135*(1), 260-264.
[http://dx.doi.org/10.1006/jmre.1998.1542] [PMID: 9799704]

[123] Darrasse, L.; Ginefri, J.C. Perspectives with cryogenic RF probes in biomedical MRI. *Biochimie,* **2003**, *85*(9), 915-937.
[http://dx.doi.org/10.1016/j.biochi.2003.09.016] [PMID: 14652180]

[124] Griffin, R.G.; Prisner, T.F. High field dynamic nuclear polarization--the renaissance. *Phys. Chem. Chem. Phys.,* **2010**, *12*(22), 5737-5740.
[http://dx.doi.org/10.1039/c0cp90019b] [PMID: 20485782]

[125] Skutnik, J.M.; Assink, R.A.; Celina, M. High-Sensitivity Chemical Derivatization NMR Analysis for Condition Monitoring of Aged Elastomers. *Polymer (Guildf.),* **2004**, *45*(22), 7463-7469.
[http://dx.doi.org/10.1016/j.polymer.2004.08.058]

[126] Mansour, F.R.; Khairy, M.A. Pharmaceutical and biomedical applications of dispersive liquid-liquid microextraction. *J. Chromatogr. B Analyt. Technol. Biomed. Life Sci.,* **2017**, *1061-1062*, 382-391.
[http://dx.doi.org/10.1016/j.jchromb.2017.07.055] [PMID: 28802218]

[127] Mansour, F.R.; Danielson, N.D. Solidification of floating organic droplet in dispersive liquid-liquid microextraction as a green analytical tool. *Talanta,* **2017**, *170*, 22-35.
[http://dx.doi.org/10.1016/j.talanta.2017.03.084] [PMID: 28501162]

[128] Mansour, F.R.; Danielson, N.D. Solvent-terminated dispersive liquid-liquid microextraction: a tutorial. *Anal. Chim. Acta,* **2018**, *1016*, 1-11.
[http://dx.doi.org/10.1016/j.aca.2018.02.005] [PMID: 29534799]

CHAPTER 6

NMR as a Tool for Exploring Protein Interactions and Dynamics

Qamar Bashir and Naeem Rashid[*]

School of Biological Sciences, University of the Punjab, Quaid-e-Azam Campus, Lahore 54590, Pakistan

Abstract: Proteins are vital players that mediate a vast majority of cellular functions. NMR spectroscopy originally developed by physicists for investigation of nuclear properties, now represents highest applications in chemistry and biochemistry. NMR has been extensively utilized by structural biologists for exploring protein-ligand interactions and by medicinal chemists for drug discovery. The ligands investigated involved small organic molecules, peptides, proteins and nucleic acids. Recently, there has been increasing interest in the dynamic studies of these protein-ligand interactions. These applications are provided by a multitude of NMR experiments ranging from the simple one-dimensional ^{1}H spectrum to complex multidimensional NMR approaches. Chemical shift perturbation analysis allows for delineation of the binding interface, determination of the dissociation constants and estimation of ligand binding kinetics. Paramagnetic Relaxation Enhancement NMR spectroscopy has been widely used to visualize the weakly populated states and describes the process of protein complex formation. These approaches have been demonstrated for substrate binding, allostery, state equilibria and macromolecular self-association. NMR spectroscopy allows for characterization of minor conformational dynamic differences in structurally similar proteins. Target Immobilized NMR screening represents another approach to drug discovery that allows ligand screening for challenging targets. NMR spectroscopy can also be applied in combination with other techniques including X-ray crystallography and various computational methods to achieve greater coverage than any of the individual methods. This chapter is focused on the applications of NMR in exploring protein-ligand interactions and dynamics.

Keywords: Chemical shift perturbation, Encounter complex, Ligand binding, NMR, Paramagnetic relaxation enhancement, Protein dynamics, Protein-ligand interaction, Spectroscopy, Specific complex, Target immobilized NMR screening.

---

[*] **Corresponding author Naeem Rashid:** School of Biological Sciences, University of the Punjab, Quaid-e-Azam Campus, Lahore 54590, Pakistan; Tel: +924299231534; E-mail: naeemrashid37@hotmail.com

Atta-ur-Rahman and M. Iqbal Choudhary (Eds.)
All rights reserved-© 2020 Bentham Science Publishers

INTRODUCTION

Protein-ligand interactions are vital for the maintenance and proper functioning of biological systems. A variety of cellular processes are carried out by proteins through interactions with multiple ligands including other proteins, small molecules and nucleic acids. These processes include signal transduction [1, 2], electron transport [3, 4], cellular metabolism [5, 6], muscle contraction [7, 8], membrane transport [9, 10], gene expression by transcription factors [11, 12], regulation of cytoskeleton [13, 14], enzymatic reactions and enzyme inhibition by intracellular inhibitors [15 - 18]. Abnormality in these interactions can lead to diseases like cancer, Alzheimer's and Creutzfeldt-Jakob disease [19, 20]. Protein-ligand interactions are the physical events, directed by the biochemical events of electrostatic forces, hydrophobic effect, hydrogen bonding, van der Waals and pi interactions. The affinity of protein-ligand interaction is a thermodynamic property described by the dissociation constant (K_d), which is ratio of the individual rate constants of dissociation (k_{off}) and association (k_{on}). The K_d values can range from 10^{-2} M to 10^{-16} M [21, 22]. Depending on the function performed, the protein-ligand interactions are tuned in terms of the strength, specificity and life time of the final complex. On one hand are the specific and static complexes of antigens and antibodies as well as enzymes and their inhibitors. Such proteins have single partners and avoid interactions with other cellular components. In such cases, strong binding is essential to lock the complexes in a single, well-defined orientation. These complexes are characterized by their low dissociation constant (10^{-15} M to 10^{-16} M), high binding energy (up to -21 kcal/mol) and long life-times even up to several days [23]. On the other hand are the transient, weak complexes involved in signal transduction and electron transport. Such events require a high turnover and fast association/dissociation of the partners. Proteins involved in these interactions may recognize multiple ligands, and a high specificity in these complexes is avoided to gain a rapid dissociation. Such complexes have high dissociation constants (μM to mM), low binding energies and short life times of millisecond time scale [24].

The study of protein-ligand interactions is important for the understanding of mechanisms underlying the cellular processes and for drug development. Protein-ligand interactions have been studied at increasing pace by a wide range of experimental techniques [25] including UV-Visible spectroscopy [26], analytical ultracentrifugation [27], microscale thermophoresis [28], surface plasmon resonance [29], isothermal titration calorimetry [30], circular dichroism [31], dynamic light scattering [32], atomic force microscopy [33], mass spectrometry [34], differential scanning fluorimetry [35], small angle X-ray scattering [36], fluorescence microscopy [37], quartz crystal microbalance [38] and NMR [39]. These methods provide information on multiple aspects of protein-ligand interactions including association, dissociation, conformational changes on binding, kinetic and thermodynamics parameters, with some limitations associated with each technique. It has become increasingly recognized that proteins and ligands do not behave as static objects in solution, rather they are dynamic bodies [40, 41]. There are different kinds of fluctuations, transitions, conformational changes, movements, bond vibrations and rotations going on. They correspond to the local fluctuations of chemical bonds, regional flexibility of residues relative to each other and global movements of protein domains. It also encompasses relative movements of proteins or ligands on or around the binding interface. In terms of structure, kinetics and thermodynamics, NMR is a versatile and powerful technique that presents site-specific information of protein-interactions. It comprises a number of experiments that allow for description of the binding interface, derivation of thermodynamic parameters and characterization of protein dynamics. This chapter highlights the subset of NMR experiments that can be utilized to explore protein-ligand interactions and dynamics.

Protein Dynamics and the Encounter Complex

Protein dynamics cover a broad range of movements within or across the protein surface. Within living organisms, proteins are in constant motion and are interacting with other biomolecules to convey biological messages. A protein must physically interact with its ligand and form a productive complex for successful execution of the assigned task. This interaction has to be very specific to allow for binding with a specific molecule, thereby avoiding interactions with other cellular components. This is achieved by the presence of a specific binding interface that allows for selective recognition of the desired ligand. This binding interface is composed of small surface patches on the protein and ligand and is small as compared to the whole protein surface. If the ligand has to find and bind to the specific interaction interface through mere diffusion-driven collisions in solution, most of the collisions will be nonproductive due to the small chance of directly hitting the small interface. However, in most biological processes such as signal transduction cascades and electron transfer reactions, a fast association of the interacting molecules is crucial. This is achieved by the formation of a dynamic encounter complex [42] that accelerates the formation of the final specific complex by increasing the number of successful collisions, even if they are not directly on target (Fig. **1**).

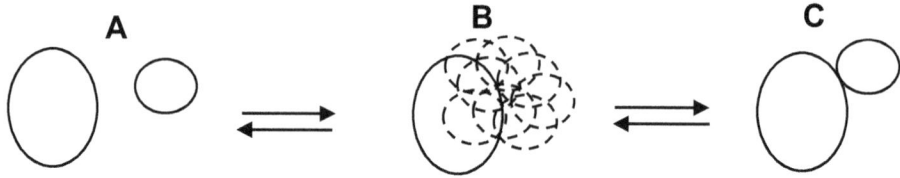

Fig. (1). Model for protein complex formation. Free proteins (**A**), the dynamic encounter complex (**B**) and well – defined specific complex (**C**).

In this view, protein complex formation is a multistep process. The interacting molecules approach each other by diffusion-controlled random collisions and associate in a non-specific manner to form an encounter complex. In the encounter complex, the molecules are kept in proximity by non-specific long-range electrostatic interactions [16, 43]. This weak association increases the life time of intermolecular proximity, thereby allowing the ligand to diffuse around the protein surface until the specific site is found. Such a two-dimensional search is faster than repeated three-dimensional collisions. The encounter complex may comprise multiple (small) ligands in multiple orientations covering the protein surface. The molecules in the encounter complex then dissociate and rebind, or reorient their surface patches to form the single orientation specific complex. The specific complex is dominated by short range interactions like hydrogen bonding and van der Waals interactions, which may or may not involve water molecules. The encounter and specific complexes exist in equilibrium and their relative population is linked to the difference in total free energy content of both complexes, not only including short and long-range protein-ligand interactions, but also including changes in water-protein interactions and changes in configurational entropy of protein and ligand in both complexes. However, it seems safe to assume that the greater the specificity of a protein-ligand complex, the larger is the relative fraction of the specific complex and *vice versa* [44 - 47]. Depending on the cellular functions, protein complexes can exist in either of the two states. A protein complex can be a purely specific or the encounter complex [48 - 52]. The relative populations of the two complexes can be shifted by mutating the interface residues [53]. The encounter complex has been detected in number of protein-protein and protein-DNA complexes [54 - 57].

Chemical Shift Perturbation Analysis

Chemical shift perturbation (CSP) analysis is a common and widely used NMR method to detect and quantify protein-ligand interactions [58 - 63]. It provides site-specific information and helps to delineate the binding interface (Fig. **2**). The technique involves recording of a heteronuclear single quantum coherence (HSQC) or transverse relaxation-optimized (TROSY) NMR spectrum of a ^{15}N labeled protein and titrating the ligand in a series of NMR experiments [64, 65]. The HSQC or TROSY spectrum is like a fingerprint of the protein. Each peak in the spectrum corresponds to a specific amide group in the protein, depending upon its chemical environment. The nuclei in the interface undergo changes in the chemical environment due to the ligand binding, which leads to chemical shift perturbations. The appearance of the NMR spectrum is determined by the exchange rate of the free and bound protein, or more precisely the dissociation-rate, k_{off}, of the complex. On the chemical shift time scale, a fast exchange regime is achieved when k_{off} is at least ten times greater than the chemical shift difference (in radians) between the free and bound states. This is typically related to the interactions with affinity (K_d) weaker than about 3 µM. During a fast exchange situation, the observed chemical shift corresponds to the weighted-population average of the free and bound forms. Slow and intermediate exchange rates occur when k_{off} is equal to or smaller than the chemical shift difference between the free and bound states. They result in line broadening and make it difficult to determine the binding constant. For fast exchange systems, the chemical shift changes upon ligand binding can be fitted to estimate the binding constant of the interaction. The average amide chemical shift perturbation $\Delta\ddot{a}_{avg}$ can be derived from Equation 1 [66 - 68].

$$\Delta\delta_{avg} = \sqrt{\frac{(\Delta\delta^N/5)^2 + (\Delta\delta^H)^2}{2}} \qquad (1)$$

Where $\Delta\ddot{a}^N$ and $\Delta\ddot{a}^H$ are the chemical shift perturbations, in ppm, of the amide nitrogen and proton, respectively.

The binding constant can be derived from the two-parameter non-linear square least fit of chemical shift perturbation against protein to ligand ratio as given by Equation 2 [66 - 68].

$$\Delta\delta = {}^1\!/_2\, \Delta\delta_0 \left[A - \sqrt{A^2 - \frac{4}{R}} \right] \qquad (2a)$$

$$A = 1 + 1/R + \frac{[P]_0 + R[L]_0}{R[P]_0[L]_0 K_B} \tag{2b}$$

$\Delta \ddot{a}$ is the chemical shift change at a given point in the titration; $\Delta \ddot{a}_0$ is the chemical shift change for 100% bound protein; R is the ratio of protein to ligand at a given titration point; $[P]_0$ is the concentration of stock solution of protein; $[L]_0$ is the starting concentration of ligand; K_B is the binding constant of the protein-ligand complex. This equation assumes that the protein concentration does not change (much) during titration. If it does, the actual protein concentration (P + PL) needs to be computed for each titration point and inserted for $[P]_0$.

The size of the chemical shift changes is likely to report on the character of protein-ligand complexes [51 - 53]. Specific complexes exist in single well-defined orientations and are dominated by short-range interactions like salt bridges, hydrogen bonding and hydrophobic effect, strongly perturbing the chemical/magnetic environment of the interface residues, leading to large chemical shift changes (Fig. 3). Encounter complexes exist as ensemble of multiple orientations of relative energy and at least a single solvent layer exists at the interface [3, 42]. Chemical shift changes are small in such complexes as they are averaged over all the orientations.

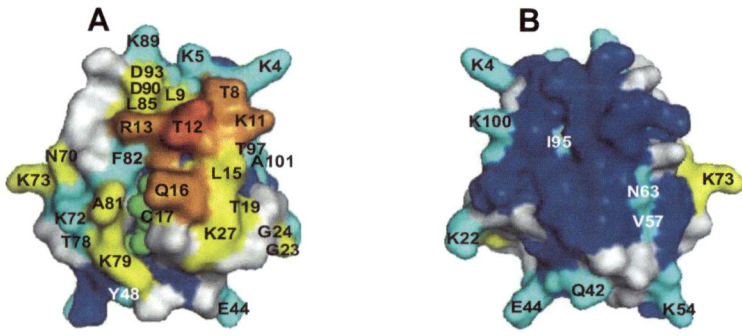

Fig. (2). Chemical shift perturbation mapping of the cytochrome c interface during it interaction with L-galactono-γ-lactone dehydrogenase. The figure shows surface representation of cytochrome c. **A** shows the front face of cytochrome c with the binding site, **B** represents the back side of the protein. The binding residues are colored according to the size of chemical shift changes. Residues with $\Delta \ddot{a}^N > 0.5$ ppm are colored in red, $\Delta \ddot{a}^N > 0.25$ ppm in orange, $\Delta \ddot{a}^N > 0.1$ ppm in yellow, $\Delta \ddot{a}^N > 0.05$ ppm in cyan. Blue colored residues do not show any chemical shift changes ($\Delta \ddot{a}^N < 0.05$ ppm). Grey color represents unassigned and the proline residues. Reprinted with permission from the authors of a study [67]. Copyright 2013 Federation of European Biochemical Societies.

Chemical shift perturbation analysis is particularly successful in exploring protein-ligand interactions on the fast exchange regime that have high dissociation rates compared to the chemical shift differences between the free and bound states. The technique results in ambiguities in case of complexes that are accompanied by large conformational changes upon ligand binding. The conformational changes may also result in chemical shift perturbations in the protein amides that may not be located in the binding site. Chemical shift perturbation is not limited to ^{15}N-^{1}H spectra. Selective ^{13}C-labelling of methyl groups and (^{13}C-^{1}H) TROSY NMR spectroscopy allows for the study of biomolecular interactions in asymmetric and large molecular complexes [69]. The ligand-induced chemical shift changes can be used as restraints to dock the ligand onto the protein. This allows for the determination of not only the position but also the ligand orientation. HADDOCK is a popular software program that makes use of the chemical shift perturbations [70] for computing conformational ensembles of the complex.

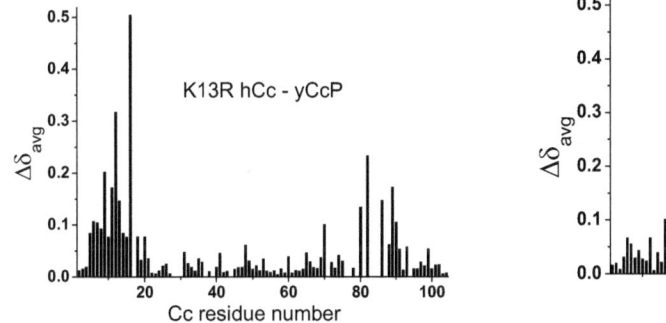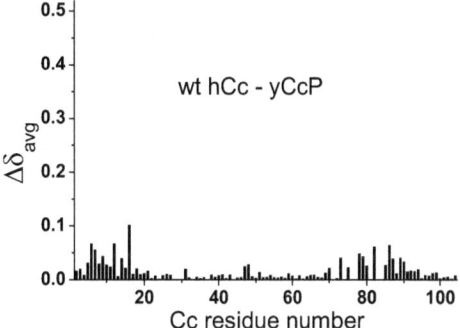

Fig. (3). The average chemical shift perturbations of the wild type and K13R horse cytochrome c in complex with the yeast cytochrome c peroxidase showing different degree of specificity. The average chemical shift perturbation is plotted against the residue number. The greater size of average chemical shift perturbations in K13R hCc - yCcP complex implies more specificity. On the other hand, smaller size of the chemical shift changes in wt hCc - yCcP represents that the complex is more diffuse. Reprinted with permission from the authors of a study [68]. Copyright 2014 Federation of European Biochemical Societies.

Target Immobilized NMR Screening

There are many NMR experiments to detect protein-ligand interactions by detecting the NMR spectra of the ligand and not the protein, such as saturation transfer difference (STD) [71], water-ligand observed via gradient spectroscopy (waterLOGSY) *etc.* [72]. Target Immobilized NMR screening (TINS) is a proprietary technique used in fragment-based drug discovery (FBDD) process for the ligand screening of more challenging targets [73, 74]. FBDD is an important tool in drug discovery to search for new lead compounds. Fragments are small organic compounds with low binding affinity to the target that can be combined and then further optimized to develop more potent lead compounds of high affinity. FBDD differs from high throughput screening (HTS) in multiple ways. HTS involves library screening of millions of lead compounds having high molecular weight (~500 Da) and dissociation constants in nano-molar range, while FBDD involves screening of thousands of small molecular weight (~200 Da) hits having low affinities in micro-molar range. Several databases with millions of chemical compounds are available for virtual drug screening, including ZINC, NCI, MDL, PubChem, ChemNavigator and CCDC's Cambridge structural database [75 - 77]. HTS is the most widely used approach that is applied by pharmaceutical companies in their drug discovery pipelines. However, HTS does not often achieve the goal of getting viable lead compounds, particularly for the challenging protein targets. Moreover, lead optimization further increases the size, complexity and lipophilicity of the drug-like compound, thereby violating Lipinski's rule of five (Ro5) [78, 79]. Ro5 states that a drug-like compound may have: no more than 5 hydrogen bond donors, no more than 10 hydrogen bond acceptors, molecular weight of less than 500 daltons and octanol-water partition coefficient no greater than 5. An orally active drug may not have more than one violation of these criteria. Favorable exceptions to Ro5 are found among natural products [80]. FBDD is more suited for the initial screening of low affinity small molecules from diverse libraries that can be further optimized into lead compounds. The lead compounds produced through FBDD are different from those selected through HTS and have higher success rates for development into new drugs [73, 81]. High resolution NMR spectroscopy is a highly validated technique for screening of the low affinity FBDD ligands and guiding the design of high affinity drug-like compounds [61, 62]. Most of the NMR methods require large quantities of readily soluble targets that sometimes limit their application in FBDD as the complex mixtures must be discarded or re-purified after each binding experiment. However, TINS is the most suited NMR method of FBDD because it allows for screening of fragment libraries of up to 10, 000 compounds to a single target sample.

As the name implies, TINS involves immobilization of the target sample on a solid support that is compatible with the NMR methods. Equimolar amounts of fragment mixtures from the library are injected into the target sample and the binding is detected by recording simple 1D NMR spectrum of the compounds. Those compounds that bind to the immobilized target have NMR resonances that are so broad that they cannot be detected anymore by solution NMR methods, and their signals disappear from the spectrum. The NMR spectrum of a sample containing the same compounds is subtracted from the target sample spectrum (Fig. **4**). The TINS spectrum then contains only the resonances from the fragments that bind to the target sample. As the target is immobilized, it can be washed clear of the low-affinity compounds after every experiment and is readily available for screening of the next mixture. Contrary to most other NMR experiments, TINS thus requires significantly less amount of target (3–5 mg) protein. Moreover, it can be applied for screening of targets that are insoluble such as membrane proteins or that are difficult to produce [73, 82]. Experiments have indicated that the binding capacity of the target is not reduced even after the application of 2000 different fragments, emphasizing its potential for screening of large fragment libraries. A mixture of 3–7 compounds can be applied in a single measurement. The sensitivity of TINS allows for monitoring the binding across a wide range of affinities (K_d from 60 to 5000 µM) and the possibility of losing the interesting hits is sparse [73]. On the other hand, the stoichiometric quantities of the target and the compound eliminate extremely weak binding at low affinity second binding sites that may exist. TINS allows for analysis of competitive binding, thereby permitting the fast characterization of the common binding site. Membrane proteins are important pharmaceutical drug targets. However, their solubilization and reconstitution in artificial membranes hinder their applications as targets in FBDD methods. Therefore, TINS offers a method of great choice for fragment screening of membrane proteins as it does not necessitate solubilization. TINS has been successfully used on cytosolic proteins, membrane proteins as well as nucleic acid targets [82].

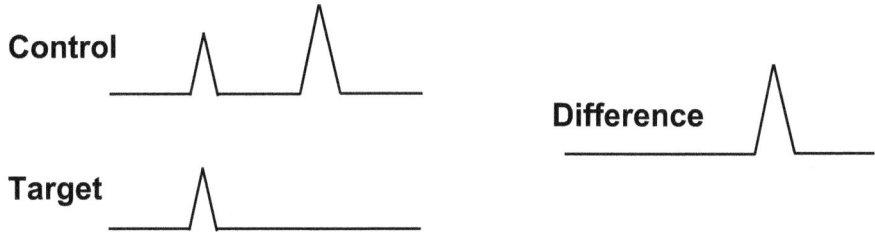

Fig. (4). Model representation of Target Immobilized NMR screening. The 1D NMR spectrum of the fragments is observed in the presence of target as well as control. The subtraction of target spectrum from the control gives the TINS spectrum. The difference spectrum contains only the resonances from binding fragments.

Paramagnetic Relaxation Enhancement

Paramagnetic relaxation enhancement (PRE) is a powerful method to study the surface dynamics in protein complexes and to visualize lowly populated states [54 - 57]. Paramagnetic relaxation of NMR nuclei originates from unpaired electrons (paramagnetic center) that have large magnetic dipoles. The effect is inversely proportional to the sixth power of distance ($1/r^6$) from the paramagnetic center. Any nucleus within the vicinity (10 Å to 35 Å) of the paramagnetic center results in paramagnetic relaxation enhancement [83]. A large number of metalloproteins have intrinsic paramagnetic centers on their cofactor metal ions. On other proteins, the paramagnetic centers can be incorporated in the form of nitroxide spin labels (Fig. 5) or tags with metal centers [54, 84 - 86]. These spin labels can be carefully attached anywhere on the protein through disulfide bonds with the introduced, by mutation, single or double cysteine residues. The spin labels attached via single cysteine residue have more mobility as compared to those attached via double cysteine residues.

Fig. (5). Covalent attachment of the nitroxide spin label on to the protein surface via cysteine residue.

The paramagnetic effects can be measured by first recording NMR spectra (HSQC or TROSY) of ^{15}N labeled protein in the presence of a diamagnetic control and then in the presence of a paramagnetic center. The ratio of peak intensities in the presence of paramagnetic and diamagnetic centers gives the quantitative paramagnetic effect. The paramagnetic relaxation enhancement can be intramolecular if the paramagnetic center and the observed nuclei exist within the same protein. The intramolecular PREs can help to investigate conformational changes within the protein, induced by ligand binding. Similarly, the opening or closing of binding cavities upon ligand/substrate interaction can also be studied [87]. The intramolecular PREs require ^{15}N labeled protein harboring the paramagnetic center and the unlabeled ligand. Intermolecular PREs arise when a paramagnetic center is attached to the protein and its effects are measured on the ligand nuclei. In this scenario, the protein carrying the paramagnetic center is unlabeled and the labeled ligand nuclei are observed in the NMR spectra. PRE has proved to be particularly important for visualization and characterization of the encounter complex. As the PRE depends on the r^{-6}, the effect is very strong on the nuclei close to the paramagnetic center, it can be used to detect even minor ligand conformations that exist for a very short period of time during complex formation. PRE has been successfully utilized to study dynamics in protein-protein complexes [55, 56], protein-DNA complexes [56, 57], state equilibria [88], allostery [89] and macromolecular self-association [90, 91]. By applying spin labels at various positions on the protein surface and observing the paramagnetic effects, the complete surface sampled by the encounter complex has been mapped (Fig. **6**) [44].

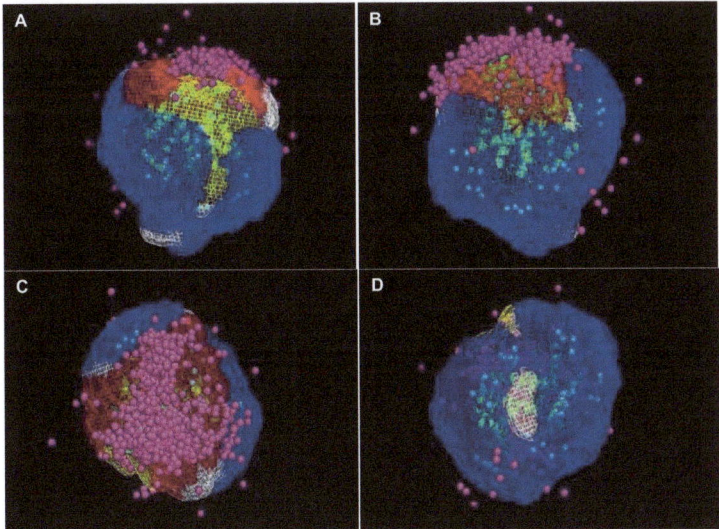

Fig. (6). Representation of the encounter complex of cytochrome c (Cc) and cytochrome c peroxidase (CcP). The proteins in the specific complex are shown as green ribbons. The grid of possible Cc positions around CcP is shown as mesh, colored according to encounter complex visualized by Paramagnetic relaxation enhancement. Blue mesh represents area around CcP that is either not sampled or visited for less than 1% of the time by Cc. Spin labels attached in this area do not show any paramagnetic effects. Red mesh represents the area that is sampled by the proteins in the encounter complex as evidenced by the strong paramagnetic effects. Yellow mesh represents the surface where attached spin labels resulted only few paramagnetic effects. White mesh represents the area where no data is available. The positions of attached spin labels are indicated by cyan spheres. To include mobility, the oxygen atom of the nitroxide spin label is represented by four cyan spheres. Magenta spheres show the simulated encounter complex. Each sphere shows the center of mass of Cc around CcP. **A, B, C, D** represent different views of the images. Reprinted with permission from reference [44]. Copyright 2010 American Chemical Society.

The quantitative paramagnetic relaxation enhancement can be calculated from the resonance intensities in the paramagnetic sample and the diamagnetic control as given in the equation 3 [92].

$$\frac{I_{para}}{I_{dia}} = \frac{R_2 exp(-t\Gamma_2)}{R_2 + \Gamma_2} \qquad (3)$$

Where I_{para} and I_{dia} are the peak intensities in the paramagnetic and diamagnetic samples. R_2 is the transverse relaxation rate, $\tilde{A}_2$ is the PRE and t is the total insensitive nuclei enhanced by polarization transfer (INEPT) evolution time of

HSQC. R_2 for each peak, can be calculated from the width at half-height ($\Delta v_{1/2}$) of a Lorentzian fit in the proton dimension by using $R_2 = \pi \Delta v_{1/2}$.

CONCLUDING REMARKS

High resolution NMR spectroscopy offers an array of methods to selectively address various aspects of protein-ligand interactions and dynamics. CSP analysis remains the popular and well-suited technique to investigate protein-ligand binding studies. It enables estimation of binding constants, depiction of binding interface and derivation of thermodynamic parameters. CSP requires ^{15}N labeling of either the protein or the ligand and then recording HSQC or TROSY spectrum in a series of protein-ligand titrations. The chemical shift changes at each titration point are measured and can be fitted against protein/ligand ratio to determine the binding parameters. The size of chemical shift changes reports on the relative population of the encounter complex. The more specific the complex, the greater the size of chemical shift perturbations and vice versa. However, CSP gives only the qualitative measure of the protein dynamics and a more sensitive technique, like PRE, has to be applied for the quantitative measurement. Due to relaxation and line broadening effects in the NMR experiments, CSP is restricted to the binding studies of protein of up to roughly 40 kDa molecular weight.

TINS has direct application in drug discovery efforts. It has been developed for the screening of compound libraries in fragment-based drug discovery process that has many advantages over the conventional methods. It is highly sensitive to record even weak binding affinities, hence minimizing the risk of losing interesting hits that can be further developed into lead compounds. TINS assay requires only a small amount of sample and allows screening of challenging targets including proteins and nucleic acids. TINS is a simple 1D spectrum that permits ligand screening of larger proteins that pose limitations for analysis by conventional NMR methods. Furthermore, TINS is successfully applied to the challenging membrane protein targets, as it does not require protein to be soluble.

PRE spectroscopy has been widely applied to study protein dynamics and visualize the encounter complex. Paramagnetic tags can be easily engineered on the protein surface via disulfide bonds. It is being utilized to map out the complete surface sampled by the encounter complex. The technique is extremely sensitive to observe minor orientations, or populations that exist for very short time. However, PRE is averaged over all the populations and does not provide information on the individual orientations. To this end, theoretical methods can play important role. The specific and the encounter complexes are governed by different types of interactions. The encounter complex is dominated by long-range interactions while the specific complex is stabilized by the short-range interactions. It is possible to simulate either of the two complexes by selectively defining these interactions in the simulation method. PRE gives the population average of the encounter and specific complexes, while the pure encounter complex can be simulated. Combination of the PRE and the simulation data gives the relative populations of the two complexes. The relative population and the surface area sampled by the encounter complex vary for different complexes depending on their biological functions. Tight protein-ligand complexes are likely to have higher population of the specific complex and weak complexes have more population of the encounter complex [42]. A protein complex can be purely specific or even purely encounter complex. The NMR methods can be supplemented by other techniques to have more coverage and accuracy.

CONSENT FOR PUBLICATION

Not applicable.

CONFLICT OF INTEREST

The authors confirm that this chapter contents have no conflict of interest.

ACKNOWLEDGEMENTS

Declared none.

REFERENCES

[1] Hemsath, L.; Dvorsky, R.; Fiegen, D.; Carlier, M-F.; Ahmadian, M.R. An electrostatic steering mechanism of Cdc42 recognition by Wiskott-Aldrich syndrome proteins. *Mol. Cell,* **2005**, *20*(2), 313-324.
[http://dx.doi.org/10.1016/j.molcel.2005.08.036] [PMID: 16246732]

[2] Marchand, J-B.; Kaiser, D.A.; Pollard, T.D.; Higgs, H.N. Interaction of WASP/Scar proteins with actin and vertebrate Arp2/3 complex. *Nat. Cell Biol.,* **2001**, *3*(1), 76-82.
[http://dx.doi.org/10.1038/35050590] [PMID: 11146629]

[3] Bashir, Q.; Scanu, S.; Ubbink, M. Dynamics in electron transfer protein complexes. *FEBS J.,* **2011**, *278*(9), 1391-1400.
[http://dx.doi.org/10.1111/j.1742-4658.2011.08062.x] [PMID: 21352493]

[4] Andrałojć, W.; Hiruma, Y.; Liu, W.M.; Ravera, E.; Nojiri, M.; Parigi, G.; Luchinat, C.; Ubbink, M. Identification of productive and futile encounters in an electron transfer protein complex. *Proc. Natl. Acad. Sci. USA,* **2017**, *114*(10), E1840-E1847.
[http://dx.doi.org/10.1073/pnas.1616813114] [PMID: 28223532]

[5] Kleppe, R.; Martinez, A.; Døskeland, S.O.; Haavik, J. The 14-3-3 proteins in regulation of cellular metabolism. *Semin. Cell Dev. Biol.,* **2011**, *22*(7), 713-719.
[http://dx.doi.org/10.1016/j.semcdb.2011.08.008] [PMID: 21888985]

[6] DeBerardinis, R.J.; Thompson, C.B. Cellular metabolism and disease: what do metabolic outliers teach us? *Cell,* **2012**, *148*(6), 1132-1144.
[http://dx.doi.org/10.1016/j.cell.2012.02.032] [PMID: 22424225]

[7] Gergely, J. Molecular aspects of muscle contraction and regulation. *Basic Res. Cardiol.,* **1977**, *72*(2-

3), 109-117.
[http://dx.doi.org/10.1007/BF01906348] [PMID: 860989]

[8]   Holmes, K.C. Muscle proteins--their actions and interactions. *Curr. Opin. Struct. Biol.,* **1996**, *6*(6), 781-789.
[http://dx.doi.org/10.1016/S0959-440X(96)80008-X] [PMID: 8994878]

[9]   Alberts, B.; Bray, D.; Lewis, J.; Raff, M.; Roberts, K.; Watson, J.D. Principles of membrane transport. In: *Molecular Biology of the Cell*, 3rd ed.; Alberts, B., Johnson, A., Lewis. J., Raff, M., Roberts, K., Wlter, P., Eds.; Garland Science: New York, **1994**; pp. 508-512.

[10] Theobald, D.L.; Miller, C. Membrane transport proteins: surprises in structural sameness. *Nat. Struct. Mol. Biol.,* **2010**, *17*(1), 2-3.
[http://dx.doi.org/10.1038/nsmb0110-2] [PMID: 20051980]

[11] Sugase, K.; Dyson, H.J.; Wright, P.E. Mechanism of coupled folding and binding of an intrinsically disordered protein. *Nature,* **2007**, *447*(7147), 1021-1025.
[http://dx.doi.org/10.1038/nature05858] [PMID: 17522630]

[12] Turjanski, A.G.; Gutkind, J.S.; Best, R.B.; Hummer, G. Binding-induced folding of a natively unstructured transcription factor. *PLOS Comput. Biol.,* **2008**, *4*(4): e1000060.
[http://dx.doi.org/10.1371/journal.pcbi.1000060] [PMID: 18404207]

[13] dos Remedios, C.G.; Chhabra, D.; Kekic, M.; Dedova, I.V.; Tsubakihara, M.; Berry, D.A.; Nosworthy, N.J. Actin binding proteins: regulation of cytoskeletal microfilaments. *Physiol. Rev.,* **2003**, *83*(2), 433-473.
[http://dx.doi.org/10.1152/physrev.00026.2002] [PMID: 12663865]

[14] Hepler, P.K. The cytoskeleton and its regulation by calcium and protons. *Plant Physiol.,* **2016**, *170*(1), 3-22.
[http://dx.doi.org/10.1104/pp.15.01506] [PMID: 26722019]

[15] Wallis, R.; Moore, G.R.; James, R.; Kleanthous, C. Protein-protein interactions in colicin E9 DNase-immunity protein complexes. 1. Diffusion-controlled association and femtomolar binding for the cognate complex. *Biochemistry,* **1995**, *34*(42), 13743-13750.
[http://dx.doi.org/10.1021/bi00042a004] [PMID: 7577966]

[16] Schreiber, G.; Fersht, A.R. Rapid, electrostatically assisted association of proteins. *Nat. Struct. Biol.,* **1996**, *3*(5), 427-431.
[http://dx.doi.org/10.1038/nsb0596-427] [PMID: 8612072]

[17] Shapiro, R.; Ruiz-Gutierrez, M.; Chen, C-Z. Analysis of the interactions of human ribonuclease inhibitor with angiogenin and ribonuclease A by mutagenesis: importance of inhibitor residues inside versus outside the C-terminal "hot spot". *J. Mol. Biol.,* **2000**, *302*(2), 497-519.
[http://dx.doi.org/10.1006/jmbi.2000.4075] [PMID: 10970748]

[18] Johnson, R.J.; McCoy, J.G.; Bingman, C.A.; Phillips, G.N., Jr; Raines, R.T. Inhibition of human pancreatic ribonuclease by the human ribonuclease inhibitor protein. *J. Mol. Biol.,* **2007**, *368*(2), 434-449.
[http://dx.doi.org/10.1016/j.jmb.2007.02.005] [PMID: 17350650]

[19] Yeger-Lotem, E.; Sharan, R. Human protein interaction networks across tissues and diseases. *Front. Genet.,* **2015**, *6*(6), 257.
[http://dx.doi.org/10.3389/fgene.2015.00257] [PMID: 26347769]

[20] Gonzalez, M.W.; Kann, M.G. Chapter 4: Protein interactions and disease. *PLOS Comput. Biol.,* **2012**, *8*(12): e1002819.
[http://dx.doi.org/10.1371/journal.pcbi.1002819] [PMID: 23300410]

[21] Lee, F.S.; Shapiro, R.; Vallee, B.L. Tight-binding inhibition of angiogenin and ribonuclease A by placental ribonuclease inhibitor. *Biochemistry,* **1989**, *28*(1), 225-230.
[http://dx.doi.org/10.1021/bi00427a031] [PMID: 2706246]

[22] Janin, J. Kinetics and thermodynamics of protein-protein interactions. *Protein protein recognition*; Klaenthous, C., Ed.; Oxford University Press: New York, **2000**, pp. 1-32.

[23] Shapiro, R.; Vallee, B.L. Interaction of human placental ribonuclease with placental ribonuclease inhibitor. *Biochemistry*, **1991**, *30*(8), 2246-2255.
[http://dx.doi.org/10.1021/bi00222a030] [PMID: 1998683]

[24] Crowley, P.B.; Ubbink, M. Close encounters of the transient kind: protein interactions in the photosynthetic redox chain investigated by NMR spectroscopy. *Acc. Chem. Res.*, **2003**, *36*(10), 723-730.
[http://dx.doi.org/10.1021/ar0200955] [PMID: 14567705]

[25] Fechner, P.; Bleher, O.; Ewald, M.; Freudenberger, K.; Furin, D.; Hilbig, U.; Kolarov, F.; Krieg, K.; Leidner, L.; Markovic, G.; Proll, G.; Pröll, F.; Rau, S.; Riedt, J.; Schwarz, B.; Weber, P.; Widmaier, J. Size does matter! Label-free detection of small molecule-protein interaction. *Anal. Bioanal. Chem.*, **2014**, *406*(17), 4033-4051.
[http://dx.doi.org/10.1007/s00216-014-7834-4] [PMID: 24817356]

[26] Nienhaus, K.; Nienhaus, G.U. Probing heme protein-ligand interactions by UV/visible absorption spectroscopy. *Methods Mol. Biol.*, **2005**, *305*, 215-242.
[http://dx.doi.org/10.1385/1-59259-912-5:215] [PMID: 15940000]

[27] Harding, S.E.; Rowe, A.J. Insight into protein-protein interactions from analytical ultracentrifugation. *Biochem. Soc. Trans.*, **2010**, *38*(4), 901-907.
[http://dx.doi.org/10.1042/BST0380901] [PMID: 20658974]

[28] Jerabek-Willemsen, M.; Wienken, C.J.; Braun, D.; Baaske, P.; Duhr, S. Molecular interaction studies using microscale thermophoresis. *Assay Drug Dev. Technol.*, **2011**, *9*(4), 342-353.
[http://dx.doi.org/10.1089/adt.2011.0380] [PMID: 21812660]

[29] Fee, C.J. Label-free, real-time interaction and adsorption analysis 1: surface plasmon resonance. *Methods Mol. Biol.*, **2013**, *996*, 287-312.
[http://dx.doi.org/10.1007/978-1-62703-354-1_17] [PMID: 23504431]

[30] Kabiri, M.; Unsworth, L.D. Application of isothermal titration calorimetry for characterizing thermodynamic parameters of biomolecular interactions: peptide self-assembly and protein adsorption case studies. *Biomacromolecules*, **2014**, *15*(10), 3463-3473.
[http://dx.doi.org/10.1021/bm5004515] [PMID: 25131962]

[31] Zsila, F.; Bikádi, Z.; Fitos, I.; Simonyi, M. Probing protein binding sites by circular dichroism spectroscopy. *Curr. Drug Discov. Technol.*, **2004**, *1*(2), 133-153.
[http://dx.doi.org/10.2174/1570163043335135] [PMID: 16472252]

[32] Lorber, B.; Fischer, F.; Bailly, M.; Roy, H.; Kern, D. Protein analysis by dynamic light scattering: methods and techniques for students. *Biochem. Mol. Biol. Educ.*, **2012**, *40*(6), 372-382.
[http://dx.doi.org/10.1002/bmb.20644] [PMID: 23166025]

[33] Whited, A.M.; Park, P.S. Atomic force microscopy: a multifaceted tool to study membrane proteins and their interactions with ligands. *Biochim. Biophys. Acta*, **2014**, *1838*(1 Pt A), 56-68.
[http://dx.doi.org/10.1016/j.bbamem.2013.04.011] [PMID: 23603221]

[34] Kool, J.; Jonker, N.; Irth, H.; Niessen, W.M. Studying protein-protein affinity and immobilized ligand-protein affinity interactions using MS-based methods. *Anal. Bioanal. Chem.*, **2011**, *401*(4), 1109-1125.
[http://dx.doi.org/10.1007/s00216-011-5207-9] [PMID: 21755271]

[35] Niesen, F.H.; Berglund, H.; Vedadi, M. The use of differential scanning fluorimetry to detect ligand interactions that promote protein stability. *Nat. Protoc.*, **2007**, *2*(9), 2212-2221.
[http://dx.doi.org/10.1038/nprot.2007.321] [PMID: 17853878]

[36] Perry, J.J.; Tainer, J.A. Developing advanced X-ray scattering methods combined with crystallography and computation. *Methods*, **2013**, *59*(3), 363-371.
[http://dx.doi.org/10.1016/j.ymeth.2013.01.005] [PMID: 23376408]

[37] Slaughter, B.D.; Li, R. Toward quantitative "*in vivo* biochemistry" with fluorescence fluctuation spectroscopy. *Mol. Biol. Cell*, **2010**, *21*(24), 4306-4311.
[http://dx.doi.org/10.1091/mbc.e10-05-0451] [PMID: 21160072]

[38] Fee, C.J. Label-free, real-time interaction and adsorption analysis 2: quartz crystal microbalance. *Methods Mol. Biol.*, **2013**, *996*, 313-322.
[http://dx.doi.org/10.1007/978-1-62703-354-1_18] [PMID: 23504432]

[39] Takeuchi, K.; Wagner, G. NMR studies of protein interactions. *Curr. Opin. Struct. Biol.*, **2006**, *16*(1), 109-117.
[http://dx.doi.org/10.1016/j.sbi.2006.01.006] [PMID: 16427776]

[40] Karplus, M.; Kuriyan, J. Molecular dynamics and protein function. *Proc. Natl. Acad. Sci. USA*, **2005**, *102*(19), 6679-6685.
[http://dx.doi.org/10.1073/pnas.0408930102] [PMID: 15870208]

[41] Karplus, M. Dynamical aspects of molecular recognition. *J. Mol. Recognit.*, **2010**, *23*(2), 102-104.
[http://dx.doi.org/10.1002/jmr.1018] [PMID: 20151415]

[42] Ubbink, M. The courtship of proteins: understanding the encounter complex. *FEBS Lett.*, **2009**, *583*(7), 1060-1066.
[http://dx.doi.org/10.1016/j.febslet.2009.02.046] [PMID: 19275897]

[43] Sheinerman, F.B.; Norel, R.; Honig, B. Electrostatic aspects of protein-protein interactions. *Curr. Opin. Struct. Biol.*, **2000**, *10*(2), 153-159.
[http://dx.doi.org/10.1016/S0959-440X(00)00065-8] [PMID: 10753808]

[44] Bashir, Q.; Volkov, A.N.; Ullmann, G.M.; Ubbink, M. Visualization of the encounter ensemble of the transient electron transfer complex of cytochrome *c* and cytochrome *c* peroxidase. *J. Am. Chem. Soc.*, **2010**, *132*(1), 241-247.
[http://dx.doi.org/10.1021/ja9064574] [PMID: 19961227]

[45] Kim, Y.C.; Tang, C.; Clore, G.M.; Hummer, G. Replica exchange simulations of transient encounter complexes in protein-protein association. *Proc. Natl. Acad. Sci. USA*, **2008**, *105*(35), 12855-12860.
[http://dx.doi.org/10.1073/pnas.0802460105] [PMID: 18728193]

[46] Vlasie, M.D.; Fernández-Busnadiego, R.; Prudêncio, M.; Ubbink, M. Conformation of pseudoazurin in the 152 kDa electron transfer complex with nitrite reductase determined by paramagnetic NMR. *J. Mol. Biol.*, **2008**, *375*(5), 1405-1415.
[http://dx.doi.org/10.1016/j.jmb.2007.11.056] [PMID: 18083191]

[47] Ubbink, M.; Ejdebäck, M.; Karlsson, B.G.; Bendall, D.S. The structure of the complex of plastocyanin and cytochrome *f*, determined by paramagnetic NMR and restrained rigid-body molecular dynamics. *Structure*, **1998**, *6*(3), 323-335.
[http://dx.doi.org/10.1016/S0969-2126(98)00035-5] [PMID: 9551554]

[48] Xu, X.; Reinle, W.; Hannemann, F.; Konarev, P.V.; Svergun, D.I.; Bernhardt, R.; Ubbink, M. Dynamics in a pure encounter complex of two proteins studied by solution scattering and paramagnetic NMR spectroscopy. *J. Am. Chem. Soc.*, **2008**, *130*(20), 6395-6403.
[http://dx.doi.org/10.1021/ja7101357] [PMID: 18439013]

[49] Liang, Z.X.; Nocek, J.M.; Huang, K.; Hayes, R.T.; Kurnikov, I.V.; Beratan, D.N.; Hoffman, B.M. Dynamic docking and electron transfer between Zn-myoglobin and cytochrome *b(*5*)*. *J. Am. Chem. Soc.*, **2002**, *124*(24), 6849-6859.
[http://dx.doi.org/10.1021/ja0127032] [PMID: 12059205]

[50] Liang, Z.X.; Kurnikov, I.V.; Nocek, J.M.; Mauk, A.G.; Beratan, D.N.; Hoffman, B.M. Dynamic docking and electron-transfer between cytochrome *b*5 and a suite of myoglobin surface-charge mutants. Introduction of a functional-docking algorithm for protein-protein complexes. *J. Am. Chem. Soc.*, **2004**, *126*(9), 2785-2798.
[http://dx.doi.org/10.1021/ja038163l] [PMID: 14995196]

[51]   Worrall, J.A.R.; Reinle, W.; Bernhardt, R.; Ubbink, M. Transient protein interactions studied by NMR spectroscopy: the case of cytochrome C and adrenodoxin. *Biochemistry,* **2003**, *42*(23), 7068-7076.
[http://dx.doi.org/10.1021/bi0342968] [PMID: 12795602]

[52]   Worrall, J.A.; Liu, Y.; Crowley, P.B.; Nocek, J.M.; Hoffman, B.M.; Ubbink, M. Myoglobin and cytochrome *b*5: a nuclear magnetic resonance study of a highly dynamic protein complex. *Biochemistry,* **2002**, *41*(39), 11721-11730.
[http://dx.doi.org/10.1021/bi026296y] [PMID: 12269814]

[53]   Volkov, A.N.; Bashir, Q.; Worrall, J.A.R.; Ullmann, G.M.; Ubbink, M. Shifting the equilibrium between the encounter state and the specific form of a protein complex by interfacial point mutations. *J. Am. Chem. Soc.,* **2010**, *132*(33), 11487-11495.
[http://dx.doi.org/10.1021/ja100867c] [PMID: 20672804]

[54]   Volkov, A.N.; Worrall, J.A.R.; Holtzmann, E.; Ubbink, M. Solution structure and dynamics of the complex between cytochrome *c* and cytochrome *c* peroxidase determined by paramagnetic NMR. *Proc. Natl. Acad. Sci. USA,* **2006**, *103*(50), 18945-18950.
[http://dx.doi.org/10.1073/pnas.0603551103] [PMID: 17146057]

[55]   Tang, C.; Iwahara, J.; Clore, G.M. Visualization of transient encounter complexes in protein-protein association. *Nature,* **2006**, *444*(7117), 383-386.
[http://dx.doi.org/10.1038/nature05201] [PMID: 17051159]

[56]   Iwahara, J.; Clore, G.M. Detecting transient intermediates in macromolecular binding by paramagnetic NMR. *Nature,* **2006**, *440*(7088), 1227-1230.
[http://dx.doi.org/10.1038/nature04673] [PMID: 16642002]

[57]   Clore, G.M. Visualizing lowly-populated regions of the free energy landscape of macromolecular complexes by paramagnetic relaxation enhancement. *Mol. Biosyst.,* **2008**, *4*(11), 1058-1069.
[http://dx.doi.org/10.1039/b810232e] [PMID: 18931781]

[58]   Zuiderweg, E.R.P. Mapping protein-protein interactions in solution by NMR spectroscopy. *Biochemistry,* **2002**, *41*(1), 1-7.
[http://dx.doi.org/10.1021/bi011870b] [PMID: 11771996]

[59]   Williamson, M.P. Using chemical shift perturbation to characterise ligand binding. *Prog. Nucl. Magn. Reson. Spectrosc.,* **2013**, *73*, 1e16.
[http://dx.doi.org/10.1016/j.pnmrs.2013.02.001]

[60]   Harner, M.J.; Mueller, L.; Robbins, K.J.; Reily, M.D. NMR in drug design. *Arch. Biochem. Biophys.,* **2017**, *628*, 132-147.
[http://dx.doi.org/10.1016/j.abb.2017.06.005] [PMID: 28619618]

[61]   Dias, D.M.; Ciulli, A. NMR approaches in structure-based lead discovery: recent developments and new frontiers for targeting multi-protein complexes. *Prog. Biophys. Mol. Biol.,* **2014**, *116*(2-3), 101-112.
[http://dx.doi.org/10.1016/j.pbiomolbio.2014.08.012] [PMID: 25175337]

[62]   Sugiki, T.; Furuita, K.; Fujiwara, T.; Kojima, C. Current NMR Techniques for Structure-Based Drug Discovery. *Molecules,* **2018**, *23*(1), 148.
[http://dx.doi.org/10.3390/molecules23010148] [PMID: 29329228]

[63]   Pellecchia, M. Solution nuclear magnetic resonance spectroscopy techniques for probing intermolecular interactions. *Chem. Biol.,* **2005**, *12*(9), 961-971.
[http://dx.doi.org/10.1016/j.chembiol.2005.08.013] [PMID: 16183020]

[64]   Bodenhausen, G.; Ruben, D.J. Natural abundance nitrogen-15 NMR by enhanced heteronuclear spectroscopy. *Chem. Phys. Lett.,* **1980**, *69*, 185-189.
[http://dx.doi.org/10.1016/0009-2614(80)80041-8]

[65]   Salzmann, M.; Pervushin, K.; Wider, G.; Senn, H.; Wüthrich, K. TROSY in triple-resonance experiments: new perspectives for sequential NMR assignment of large proteins. *Proc. Natl. Acad. Sci.*

USA, **1998**, *95*(23), 13585-13590.
[http://dx.doi.org/10.1073/pnas.95.23.13585] [PMID: 9811843]

[66] Volkov, A.N.; Bashir, Q.; Worrall, J.A.; Ubbink, M. Binding hot spot in the weak protein complex of physiological redox partners yeast cytochrome C and cytochrome C peroxidase. *J. Mol. Biol.,* **2009**, *385*(3), 1003-1013.
[http://dx.doi.org/10.1016/j.jmb.2008.10.091] [PMID: 19026661]

[67] Hervás, M.; Bashir, Q.; Leferink, N.G.; Ferreira, P.; Moreno-Beltrán, B.; Westphal, A.H.; Díaz-Moreno, I.; Medina, M.; de la Rosa, M.A.; Ubbink, M.; Navarro, J.A.; van Berkel, W.J. Communication between (L)-galactono-1,4-lactone dehydrogenase and cytochrome c. *FEBS J.,* **2013**, *280*(8), 1830-1840.
[http://dx.doi.org/10.1111/febs.12207] [PMID: 23438074]

[68] Bashir, Q.; Meulenbroek, E.M.; Pannu, N.S.; Ubbink, M. Engineering specificity in a dynamic protein complex with a single conserved mutation. *FEBS J.,* **2014**, *281*(21), 4892-4905.
[http://dx.doi.org/10.1111/febs.13028] [PMID: 25180929]

[69] Wiesner, S.; Sprangers, R. Methyl groups as NMR probes for biomolecular interactions. *Curr. Opin. Struct. Biol.,* **2015**, *35*, 60-67.
[http://dx.doi.org/10.1016/j.sbi.2015.08.010] [PMID: 26407236]

[70] Dominguez, C.; Boelens, R.; Bonvin, A.M.J.J. HADDOCK: a protein-protein docking approach based on biochemical or biophysical information. *J. Am. Chem. Soc.,* **2003**, *125*(7), 1731-1737.
[http://dx.doi.org/10.1021/ja026939x] [PMID: 12580598]

[71] Viegas, A.; Manso, J.; Nobrega, F.L.; Cabrita, E.J. Saturation-transfer difference (STD) NMR: A simple and fast method for ligand screening and characterization of protein binding. *J. Chem. Educ.,* **2011**, *88*, 990-994.
[http://dx.doi.org/10.1021/ed101169t]

[72] Huang, R.; Bonnichon, A.; Claridge, T.D.; Leung, I.K. Protein-ligand binding affinity determination by the waterLOGSY method: An optimised approach considering ligand rebinding. *Sci. Rep.,* **2017**, *7*, 43727.
[http://dx.doi.org/10.1038/srep43727] [PMID: 28256624]

[73] Vanwetswinkel, S.; Heetebrij, R.J.; van Duynhoven, J.; Hollander, J.G.; Filippov, D.V.; Hajduk, P.J.; Siegal, G. TINS, target immobilized NMR screening: an efficient and sensitive method for ligand discovery. *Chem. Biol.,* **2005**, *12*(2), 207-216.
[http://dx.doi.org/10.1016/j.chembiol.2004.12.004] [PMID: 15734648]

[74] Kobayashi, M.; Retra, K.; Figaroa, F.; Hollander, J.G.; Ab, E.; Heetebrij, R.J.; Irth, H.; Siegal, G. Target immobilization as a strategy for NMR-based fragment screening: comparison of TINS, STD, and SPR for fragment hit identification. *J. Biomol. Screen.,* **2010**, *15*(8), 978-989.
[http://dx.doi.org/10.1177/1087057110375614] [PMID: 20817886]

[75] Singh, M.; Tam, B.; Akabayov, B. NMR-Fragment based virtual screening: A brief overview. *Molecules,* **2018**, *23*(2): E233.
[http://dx.doi.org/10.3390/molecules23020233] [PMID: 29370102]

[76] Irwin, J.J.; Shoichet, B.K. ZINC--a free database of commercially available compounds for virtual screening. *J. Chem. Inf. Model.,* **2005**, *45*(1), 177-182.
[http://dx.doi.org/10.1021/ci049714+] [PMID: 15667143]

[77] Milne, G.W.; Nicklaus, M.C.; Driscoll, J.S.; Wang, S.; Zaharevitz, D. National cancer institute drug information system 3D database. *J. Chem. Inf. Comput. Sci.,* **1994**, *34*(5), 1219-1224.
[http://dx.doi.org/10.1021/ci00021a032] [PMID: 7962217]

[78] Lipinski, C.A. Drug-like properties and the causes of poor solubility and poor permeability. *J. Pharmacol. Toxicol. Methods,* **2000**, *44*(1), 235-249.
[http://dx.doi.org/10.1016/S1056-8719(00)00107-6] [PMID: 11274893]

[79] Lipinski, C.A.; Lombardo, F.; Dominy, B.W.; Feeney, P.J. Experimental and computational approaches to estimate solubility and permeability in drug discovery and development settings. *Adv. Drug Deliv. Rev.,* **2001**, *46*(1-3), 3-26.
[http://dx.doi.org/10.1016/S0169-409X(00)00129-0] [PMID: 11259830]

[80] Lipinski, C.A. Rule of five in 2015 and beyond: Target and ligand structural limitations, ligand chemistry structure and drug discovery project decisions. *Adv. Drug Deliv. Rev.,* **2016**, *101*, 34-41.
[http://dx.doi.org/10.1016/j.addr.2016.04.029] [PMID: 27154268]

[81] Keserü, G.M.; Makara, G.M. The influence of lead discovery strategies on the properties of drug candidates. *Nat. Rev. Drug Discov.,* **2009**, *8*(3), 203-212.
[http://dx.doi.org/10.1038/nrd2796] [PMID: 19247303]

[82] Früh, V.; Zhou, Y.; Chen, D.; Loch, C.; Ab, E.; Grinkova, Y.N.; Verheij, H.; Sligar, S.G.; Bushweller, J.H.; Siegal, G. Application of fragment-based drug discovery to membrane proteins: identification of ligands of the integral membrane enzyme DsbB. *Chem. Biol.,* **2010**, *17*(8), 881-891.
[http://dx.doi.org/10.1016/j.chembiol.2010.06.011] [PMID: 20797617]

[83] Iwahara, J.; Schwieters, C.D.; Clore, G.M. Ensemble approach for NMR structure refinement against (1)H paramagnetic relaxation enhancement data arising from a flexible paramagnetic group attached to a macromolecule. *J. Am. Chem. Soc.,* **2004**, *126*(18), 5879-5896.
[http://dx.doi.org/10.1021/ja031580d] [PMID: 15125681]

[84] Miao, Q.; Liu, W.M.; Kock, T.; Blok, A.; Timmer, M.; Overhand, M.; Ubbink, M. A Double-Armed, Hydrophilic Transition Metal Complex as a Paramagnetic NMR Probe. *Angew. Chem. Int. Ed. Engl.,* **2019**, *58*(37), 13093-13100.
[http://dx.doi.org/10.1002/anie.201906049] [PMID: 31314159]

[85] Liu, W.M.; Keizers, P.H.; Hass, M.A.; Blok, A.; Timmer, M.; Sarris, A.J.; Overhand, M.; Ubbink, M. A pH-sensitive, colorful, lanthanide-chelating paramagnetic NMR probe. *J. Am. Chem. Soc.,* **2012**, *134*(41), 17306-17313.
[http://dx.doi.org/10.1021/ja307824e] [PMID: 22994925]

[86] Keizers, P.H.; Ubbink, M. Paramagnetic tagging for protein structure and dynamics analysis. *Prog. Nucl. Magn. Reson. Spectrosc.,* **2011**, *58*(1-2), 88-96.
[http://dx.doi.org/10.1016/j.pnmrs.2010.08.001] [PMID: 21241885]

[87] Skinner, S.P.; Liu, W.M.; Hiruma, Y.; Timmer, M.; Blok, A.; Hass, M.A.; Ubbink, M. Delicate conformational balance of the redox enzyme cytochrome P450cam. *Proc. Natl. Acad. Sci. USA,* **2015**, *112*(29), 9022-9027.
[http://dx.doi.org/10.1073/pnas.1502351112] [PMID: 26130807]

[88] Henzler-Wildman, K.A.; Thai, V.; Lei, M.; Ott, M.; Wolf-Watz, M.; Fenn, T.; Pozharski, E.; Wilson, M.A.; Petsko, G.A.; Karplus, M.; Hübner, C.G.; Kern, D. Intrinsic motions along an enzymatic reaction trajectory. *Nature,* **2007**, *450*(7171), 838-844.
[http://dx.doi.org/10.1038/nature06410] [PMID: 18026086]

[89] Tang, C.; Schwieters, C.D.; Clore, G.M. Open-to-closed transition in apo maltose-binding protein observed by paramagnetic NMR. *Nature,* **2007**, *449*(7165), 1078-1082.
[http://dx.doi.org/10.1038/nature06232] [PMID: 17960247]

[90] Tang, C.; Ghirlando, R.; Clore, G.M. Visualization of transient ultra-weak protein self-association in solution using paramagnetic relaxation enhancement. *J. Am. Chem. Soc.,* **2008**, *130*(12), 4048-4056.
[http://dx.doi.org/10.1021/ja710493m] [PMID: 18314985]

[91] Tang, C.; Louis, J.M.; Aniana, A.; Suh, J-Y.; Clore, G.M. Visualizing transient events in amino-terminal autoprocessing of HIV-1 protease. *Nature,* **2008**, *455*(7213), 693-696.
[http://dx.doi.org/10.1038/nature07342] [PMID: 18833280]

[92] Battiste, J.L.; Wagner, G. Utilization of site-directed spin labeling and high-resolution heteronuclear nuclear magnetic resonance for global fold determination of large proteins with limited nuclear overhauser effect data. *Biochemistry,* **2000**, *39*(18), 5355-5365.
[http://dx.doi.org/10.1021/bi000060h] [PMID: 10820006]

SUBJECT INDEX

A

Acetate 25, 36, 38, 74, 86, 97, 98, 100, 101, 102, 105, 108
   formation 86
Acetic acid 12, 21, 72
Activity, inhibiting enzyme 43
AC-vector test 28
Adenocarcinoma 64, 99, 100
Adenosine receptor antagonist 26
Alcohol use disorder (AUD) 69, 70, 71, 72
   identification test (AUDIT) 69
Altered pyruvate metabolism 74
Alzheimer's disease 26, 73, 74
Amino acids 94, 97, 99
   branched-chain 94
Analysis 2, 8, 12, 21, 35, 48, 63, 64, 70, 87, 91, 101, 109
   chemometric 35
   computational 48
   histopathological 63
   metabolic 101
   multivariate 64, 70, 87, 109
   proteomic 91
   spectral 2, 8, 12, 21
   spectrometric 13
Analytes 82, 84, 110
Analytical ultracentrifugation 122
Anomeric isomers 1, 2, 12, 21
Aptamers 49, 52, 53
   binding mechanisms of 49, 53
   mutated 53
   structured ligand-free 52
Atomic force microscopy 122

B

Bacillus cereus fluoride riboswitch 51, 52
   aptamer 52
Bacillus subtilis cell lysate, Prepared 52
Background of lung cancer 63
Benign prostatic hyperplasic (BPH) 94
Benzoic acid 65

Beverages 1, 2, 3, 5, 6, 7, 8, 9, 12, 13, 18, 19, 20, 21, 25, 26, 29, 35
   coffee milk 6, 19
   commercial 6
   dairy 18
   prepared dairy 7
   simple treatment of 1, 21
   widely-consumed 26
   wine 5
Binding 43, 46, 49, 52, 123, 128, 129
   blocking receptor 43
   competitive 129
   fluoride 52
   forces, efficient 49
Binding mechanisms 49, 53
   high-affinity 49
Biological 71, 133
   functions 133
   magnetic resonance data bank 71
Biomarkers 61, 62, 65, 69, 70, 71, 72, 73, 82, 90, 93, 94, 95, 96, 107
   accurate 106, 107
   discriminatory 70
   disease 62
   prognostic 82
Biopsies 63, 81, 87, 88, 89, 104
   invasive 104
   prostate 93
Biosensors 42, 48
   novel 48
Biosynthesis, bile acid 101
Bladder stone 106, 107
Bradykinesia 73
Brain 63, 74, 81, 82, 87, 88
   biopsies 87
   metastases 87
   tumors 82, 87, 88
Breast cancer (BC) 81, 88, 89, 90
Breath condensate, exhaled 96, 99
Brown 6, 7, 19, 20
   rice liquid 6, 19
   sugar paste 7, 20

C

Caffeine 25, 26, 36, 38
  role of 26, 38
Caffeoylquinic acids 35
Cancer 7, 61, 63, 65, 68, 81, 84, 88, 93, 95, 101, 102, 103, 106, 122
  aggressive 95
  bladder 81, 106
  brain 81
  colon 101, 103
  colorectal 65, 81, 101
  lethal 106
  liver 65
  rectal 102, 103
Capping 46
  biotin-streptavidin 46
Capping of aptamers 46
Carboxylic acids 35
Carcinogens 63
Cell lung carcinoma, small 99
Cellular metabolic pathways 93
CEST-NMR technique 52
Chemical exchange saturation transfer (CEST) 51
Chemical shift 121, 124, 125, 126, 127, 132
  changes 125, 126, 127, 132
  perturbation (CSP) 121, 124, 125, 126, 127, 132
  Perturbation analysis 121, 124, 127
  perturbation mapping 126
Chemometric data treatment 25, 31
Chlorogenic acids 25, 26, 35, 36, 38
Chromatography 2, 11, 38, 62
  gas 38, 62
  high-performance anion-exchange 2
Chromophore 20
Chronic obstructive pulmonary disease (COPD) 65, 101
Circulating tumor cells (CTCs) 93
Citric acid 71
Coffea 26
  arabica 26
  canephora 26
Coffee 8, 11, 25, 26, 27, 28, 29, 30, 34, 35, 36, 38, 65
  beverages 25, 27, 29, 35, 36, 38
  bioactivities 26
  consumption 26, 28, 65
  drinking habits 26
  green 27
  ground 29, 30
  ingestion 26
  instant 8, 11
  metabolites 26, 27, 35
  roasted 27
  tested 34
Coffee beans 27, 29
  roasted 27
Collagen powder 8, 11
Collisions 123, 124
  diffusion-controlled random 124
Commercial coffees 28
Computerized tomography (CT) 81
Computing conformational ensembles 127
Concentrations 94, 105
  fecal 105
  spermine 94
Conditions 85, 98
  benign respiratory 98
  ionization 85
Consumption 72, 110
  organic solvent 110
Contaminants 84, 95
Creutzfeldt-Jakob disease 122
CSP analysis 132
Cystoscopy 106
  periodic 106

D

Definitive diagnosis of lung cancer 63
Detection 2, 11, 64, 96, 101, 103
  fluorescence 11
  modalities 101
  pulsed amperometric 2
Detector 20
  fluorescence 20
Deuterium 3, 16
  solution 3
  solvents 16
Diabetes mellitus 26
Diagnosis of lung cancer 63
Diagnosis of Parkinson's disease 73
Dicaffeoylquinic acids 35
Diseases 26, 61, 62, 63, 68, 70, 73, 74, 76, 81, 82, 84, 104, 122
  cardiovascular 26, 61, 76
  neurological 61, 74
  transmitted 70

Disorders 61, 63, 69, 73
  neurological 61, 73
  psychiatric 61
Distinct metabolite patterns 93
DMSO, cost-effective reagent 12
DNA 45, 49, 42, 43, 51
  aptamers 45, 49
    modified 51
  single-stranded 42, 43
Drinking 8, 29, 32, 33, 72
  chronic alcohol 72
  coffee 29, 32, 33
  non-chronic 72
Drinks, sport 6, 18, 19
Dynamic 85, 109, 122
    light scattering 122
    nuclear polarization 85, 109

E

Early 64, 81, 89
  detection of lymph node metastasis 89
  diagnosis of cancer 81
  distant metastasis 64
Effects 26, 52, 62, 122, 126, 130
  hydrophobic 122, 126
  influence matrix 62
  negligible structural 52
  psychostimulant 26
  quantitative paramagnetic 130
Electron transfer reactions 123
Electrophoresis 11
Electrostatic forces 122
Endometrial cancer (EC) 91, 93
Endometriosis 92
Enzyme 46, 122
  cofactors 46
  inhibition 122
Epithelial 81, 91, 92, 93
  ovarian cancer (EOC) 81, 91, 92
  -to-mesenchymal transition 93
Epithelial cells 64, 94
  normal prostatic 94
Estrogen receptor 89
Ethanol 71, 72
  metabolism 71, 72
  oxidation 72
Exhaled breath condensate (EBC) 96, 98, 99
Exonucleases 46
Extraction 27, 30, 62, 74, 110

  solid-phase 62

F

Factors 63, 69, 89, 103
  decisive prognostic 103
  extrinsic 69
  hereditary 63
  intrinsic 69
  prognostic 89
FBDD methods 129
Fecal occult blood test (FOBT) 101, 105
Fermented rice wine 6, 19
Feruloylquinic acids 35
Fingerprint 27, 76, 125
  metabolomic 76
Flavor 7, 18, 20, 26, 33, 35
  cream coconut 7, 20
  egg 7, 20
  fish 7, 20
  pleasant 26
  roasted coffee 35
  squid 18, 20
Fluorescence microscopy 122
Formic acid 108
Fragment-based drug discovery (FBDD) 127, 128, 132
Free induction decays (FIDs) 31
Function 27, 42, 65, 74, 122
  baseline lung 65
  mental 27
  mitochondrial 74

G

Gas chromatography (GC) 38, 62
Genetic 46, 62, 73
  modification 62
  regulation 46
  susceptibility 73
Glioblastoma multiforme 88
Glucuronic acid 12
Glutathione metabolism 101
Glycemic index (GI) 2
Grape juice 6, 19
Green coffee beans 28
Guanosine binding 52
Gut microbial-host co-metabolism 105

H

Heteronuclear single quantum coherence (HSQC) 71, 125, 130, 132
High-fructose corn syrup (HFCS) 1
High performance liquid chromatography (HPLC) 2, 12
High throughput screening (HTS) 128
Honey-plum vinegar 6, 18
HPLC analysis 12, 20
Human 71, 95, 96
    expressed prostatic secretions 95, 96
    metabolome database (HMDB) 71

I

Immune system 43
Infectious diseases 61
Inflammatory responses 73
Interactions 26, 38, 48, 49, 50, 52, 73, 122, 123, 124, 125, 126, 127, 132, 133
    biomolecular 127
    complex genetic environmental 73
    discriminative intermolecular 50
    water-protein 124
Invasive nature 106
Ionization suppression 62
Ion suppression 85
Isothermal titration calorimetry (ITC) 49, 122

L

Lactic 66, 71
    acid 66, 71
    acidosis 71
Leigh syndrome 74
Ligand binding 42, 52, 121
    affinities 52
    kinetics 121
    mechanism 42
Ligand-bound solution structures 50
Ligand nuclei 130
Ligands 52, 128
    cognate 52
    low affinity FBDD 128
Lipophilicity 128
Liquid chromatography (LC) 2, 62, 96, 97, 98, 100
    high performance 2

Liver cirrhosis 70
Low-density lipoproteins (LDL) 76, 88, 97
Lung cancer 63, 64, 65, 68, 69, 81, 96, 99, 101
    metabolism 64
Lymph node metastasis 89, 103

M

Magnetic dipoles 62
Magnetic resonance 63, 81, 82, 84, 93, 96
    imaging (MRI) 63, 81, 82, 84, 85, 96
    spectroscopy (MRS) 82, 85
Malignancy 81, 88, 93
    common gynecologic 93
Mestrenova software 31
Metabolic 98, 103, 105, 107
    biomarkers 98, 105, 107
    fingerprinting 103
Metabolic profiles 86, 87, 95, 97, 99, 105, 106
    blood-based 99
Metabolites 25, 27, 38, 62, 65, 68, 70, 71, 72, 74, 76, 81, 85, 87, 88, 89, 92, 93, 95, 99, 100, 101, 106, 109
    choline-containing 89
    determination and statistical analysis 70, 74
    endogenous 72
    signature 81, 87, 88
    trigonelline 65
Metabolomics 26, 27, 61, 62, 69, 72, 83, 84, 97, 109
    based disease diagnosis 62
    plasma-based 97
    urinary-based 97
Metal ions, cofactor 130
Metalloproteins 129
Michigan alcoholism screening test (MAST) 69
Microscale thermophoresis 122
Monte Carlo cross validation (MCCV) 65
Mucin production 64
Mucosa 102, 104, 105, 108
    colonic 105
    healthy 102
    oral 108
Muffin seasoning 18
Multivariate data analysis 61, 97, 99, 100
Muscle contraction 122
Mushrooms seasoning 7, 20

Subject Index

N

NAIM derivatization and NMR 2, 12
   analysis 2
   spectrometric data of aldo-sugars 12
National Institutes of Health (NIH) 70
Natural 42, 128
   modulators 42
   products 128
Neomycin 50, 52
Neurodegenerative disorders 73, 74, 76
Nicotinic acid 26
NMR 3, 15, 31, 36, 49, 50, 64, 90, 100
   fingerprinting 64
   Mestrenova software 31
   metabolite bioprofiling 100
   of RNA 49
   processing and statistical analysis 3, 15
   profiles of coffee beverages 36
   serum metabolomic signature 90
   and fluorescence spectroscopy 50
NMR-based 62, 104, 106
   analysis of serum samples 104
   medical diagnosis 62
   metabolic analysis 106
NMR-based metabolomics 63 69, 82, 89, 90, 91, 92, 105, 108
   applications 63
   analysis 90
NMR spectra 87, 96
   metabolic 87
   representative high-resolution 96
NMR spectra 72, 87
   of cerebral gliomas 87
   of social drinker and control 72
NMR spectroscopic analysis 101, 106
   of serum 101
NMR Spectroscopy 83, 121
   paramagnetic relaxation enhancement 121
   used in cancer diagnosis 83
Non-small cell lung carcinoma (NSCLC) 99, 101
Nuclear magnetic resonance 1, 25, 27, 61, 62, 70, 76, 81, 82
   metabolomics 76
   spectroscopy 61
Nucleases, vaginal 45
Nucleic acid aptamers 42
Nucleotides, guanine-based 53
Nutrition facts 7
Nutritionists 8

O

Ochratoxin 51
Oligonucleotides 43, 44
Operator intervention 109
Oral 81, 107, 108
   cancer (OC) 81, 107
   leukoplakia 108
   squamous cell carcinoma (OSCC) 108
Orthogonal partial least square 61, 70
   analyses 61
   discriminant analysis 70
OSCC development 108
Ovarian cancer 81, 91, 92, 93
   epithelial 81, 91, 92
Ovarian tumor 91, 92
   malignant 91
Over-expressed inositol 90
Oxidative condensation 15
Oxygen atom 131

P

Paramagnetic 121, 129, 130, 131
   effects 130, 131
   relaxation enhancement 121, 129, 130, 131
     quantitative 131
Parkinson's disease (PD) 26, 61, 73, 74, 76
Partial least squares (PLS) 25, 31, 61
Pathogenesis 63
Phosphothionate linkage 46
Plasma 61, 70, 72, 74, 76, 82, 87, 96, 97, 109
   metabolites studies 72
PLS regression 26, 27, 31, 32, 36
   and VIP scores 27
   model 32
Polymerase chain reaction (PCR) 43
Positron emission tomography 96
Post-SELEX 44
   modifications of aptamers 44
   optimization strategies 44
PRE spectroscopy 132
Prion protein 51
   bovine 51
Prognosis 64, 81, 95, 96, 99
   worst 64
Prognostication 104

Prognostic classification capabilities 103
Proline 87, 88, 100, 126
   residues 126
Prostate 81, 94
   abnormal 94
Prostate cancer 81, 93, 94
   growth 94
Prostate 93
   -specific antigen (PSA) 93
Prostate tumor 93, 95
   heterogeneity 93
Prostatic fluids 95
Protein 99, 121, 123, 124, 127, 129, 132
   amides 127
   catabolism 99
   complexes 124, 129
   domains 123
   dynamics 121, 123, 132
Proton 13, 15, 20, 21, 30, 89, 92
   NMR analysis 92
   NMR frequency 30
   NMR methods 89
   signals 13, 15, 20, 21
Pyrogallic acid 26

Q

Quantification of sugars 4, 17
Quantities, stoichiometric 129
Quantum mechanic analysis 52
Quartz crystal microbalance 122
Quinic acids 35, 36
Quinolinic acid 26

R

Random forest (RF) 90
Reaction 12, 43
   condensation 12
   polymerase chain 43
Refractive index (RI) 20
Regression 4, 25, 31, 70
   analysis 4
   multivariate logistic 70
Resistance 45, 46
   higher 46
   nuclease 45, 46
Resonances 36, 49, 74, 122, 128, 129
   chemical 36

surface plasmon 49, 122
Riboswitches 42, 46, 47, 48, 49, 50, 51, 52, 53
   adenine 49
   artificial 42, 48
   deoxyguanosine 52
   deoxyguanosine binding 52
   sensing 50, 52
   synthetic 53
Riboswitch regulation mechanism 47
RNA 43, 46
   oligonucleotides 43
   sequences ranging 46
RNA-based 42, 46
   intracellular sensors 46
   riboswitches 42

S

Saturation transfer difference (STD) 127
Serum 90
   metabolites 90
   metabolomic analysis 90
Short-chain fatty acids 105
Sigmoidoscopy 101
Signal(s) 1, 5, 8, 9, 14, 15, 28, 30, 31, 35, 52, 65, 68, 84, 86, 87
   overlapping 1, 8
   residual water 30, 31
   suppression 35
   transcription termination 52
Signal transduction 122, 123
   cascades 123
Small 49, 64
   -angle X-ray spectroscopy 49
   -cell lung carcinoma (SCLC) 64
Smoking 63, 64, 68, 69, 97
   cigarette 64
Soft independent modelling 70
Software 4, 50 62, 70, 74, 127
   chemometrics 62
   program, popular 127
Solution NMR 50, 128
   data 50
   methods 128
Soy sauce paste 7, 18, 20
Spectrometry 13, 14, 21
Spectroscopy 1, 3, 25, 27, 61, 62, 81, 82, 90, 94, 95, 100
Squamous cell 64, 99, 100
   carcinoma 64, 99

Status 89, 94
  hormonal receptor 89
  pathologic 94
Stimuli 27, 33, 62
  dispersed 27
  pathophysiological 62
Stress 73, 86
  oxidative 73
Structures 2, 45, 48, 51, 52, 53, 123
  deoxyguanosine-bound X-ray 53
  high-resolution 48
  identical tertiary 52
  ligand-bound 48
  non-bound conformational 48
Succinate 66, 74
Sugar-NAIM derivatives 2, 9, 12, 13, 14, 16
  preparation of 12, 16
  procedure for preparation of 12, 16
Sugars 1, 2, 4, 5, 6, 7, 8, 9, 12, 15, 16, 17, 18, 20, 21, 45
  excessive 8
  integration area of 4, 17
  native 1, 18
  non-reducing 15
  reducing 12, 18, 21
Surface plasmon resonance (SPR) 49, 122
Surgical 82, 104
  diagnostic procedure 82
  resection 104
Survival rates 64, 87, 89, 91, 107
Synthetic neomycin riboswitch 50
Syrup 1, 11, 18
  high-fructose corn 1
  maple sugar 8, 11

T

TADIM tests 25, 28
Tannic acid 26
Target(s) 121, 127, 128, 129, 132
  challenging membrane protein 132
  challenging protein 128
  immobilized NMR screening (TINS) 121, 127, 128, 129, 132
  nucleic acid 129
  pharmaceutical drug 129
Taurocholic acid 101, 105
Tests 101, 105
  blood-based DNA 101
  fecal-based DNA 101
  fecal immunochemical 101
  guaiac-fecal occult blood 105
  non-invasive diagnostic 101
Therapeutics 42, 43
  next-generation 42
Thermodynamic property 122
Thrombin 51
  binding aptamer (TBA) 51
  conjugated DNA 51
Thyroid-stimulating hormone 45
TINS 128, 129
  assay 132
  spectrum 128, 129
Tissue biopsies 102
Toxicity response 51
Transcription factors 122
Transformation, malignant 87, 108
Tumor(s) 63, 87, 89, 94, 97, 99, 100, 104, 105, 108
  colorectal 105
  lung 97
  tissue 100, 104, 105

U

Ultrasonography 81
Unified Parkinson's disease rating scale (UPDRS) 73
Uridine nucleotides 99
Urinary 106
  bladder cancer (UBC) 106
  tract infection 106

V

Variable importance in projection (VIP) 25, 32
Very low-density lipoproteins (VLDL) 97

W

Warburg effect 108
World health organization (WHO) 2, 7, 8

X

X-ray 48, 81, 121
  crystallography 48, 121
  imaging 81